Current and Evolving Practices in the Quality Control of Cosmetics

Current and Evolving Practices in the Quality Control of Cosmetics

Editor

Kalliopi Dodou

MDPI • Basel • Beijing • Wuhan • Barcelona • Belgrade • Manchester • Tokyo • Cluj • Tianjin

Editor
Kalliopi Dodou
University of Sunderland
UK

Editorial Office
MDPI
St. Alban-Anlage 66
4052 Basel, Switzerland

This is a reprint of articles from the Special Issue published online in the open access journal *Cosmetics* (ISSN 2079-9284) (available at: http://www.mdpi.com).

For citation purposes, cite each article independently as indicated on the article page online and as indicated below:

LastName, A.A.; LastName, B.B.; LastName, C.C. Article Title. *Journal Name* **Year**, *Volume Number*, Page Range.

ISBN 978-3-0365-2486-3 (Hbk)
ISBN 978-3-0365-2487-0 (PDF)

Cover image courtesy of Kalliopi Dodou

© 2021 by the authors. Articles in this book are Open Access and distributed under the Creative Commons Attribution (CC BY) license, which allows users to download, copy and build upon published articles, as long as the author and publisher are properly credited, which ensures maximum dissemination and a wider impact of our publications.
The book as a whole is distributed by MDPI under the terms and conditions of the Creative Commons license CC BY-NC-ND.

Contents

About the Editor . . . . **vii**

Kalliopi Dodou
Special Issue "Current and Evolving Practices in the Quality Control of Cosmetics"
Reprinted from: *Cosmetics* **2021**, *8*, 100, doi:10.3390/cosmetics8040100 . . . . **1**

Deborah Adefunke Adejokun and Kalliopi Dodou
A Novel Quality Control Method for the Determination of the Refractive Index of Oil-in-Water Creams and Its Correlation with Skin Hydration
Reprinted from: *Cosmetics* **2021**, *8*, 74, doi:10.3390/cosmetics8030074 . . . . **3**

Žane Temova Rakuša and Robert Roškar
Quality Control of Vitamins A and E and Coenzyme Q10 in Commercial Anti-Ageing Cosmetic Products
Reprinted from: *Cosmetics* **2021**, *8*, 61, doi:10.3390/cosmetics8030061 . . . . **9**

Foteini Biskanaki, Efstathios Rallis, George Skouras, Anastasios Stofas, Eirini Thymara, Nikolaos Kavantzas, Andreas C. Lazaris and Vasiliki Kefala
Impact of Solar Ultraviolet Radiation in the Expression of Type I Collagen in the Dermis
Reprinted from: *Cosmetics* **2021**, *8*, 46, doi:10.3390/cosmetics8020046 . . . . **27**

Manon Barthe, Clarisse Bavoux, Francis Finot, Isabelle Mouche, Corina Cuceu-Petrenci, Andy Forreryd, Anna Chérouvrier Hansson, Henrik Johansson, Gregory F. Lemkine, Jean-Paul Thénot and Hanan Osman-Ponchet
Safety Testing of Cosmetic Products: Overview of Established Methods and New Approach Methodologies (NAMs)
Reprinted from: *Cosmetics* **2021**, *8*, 50, doi:10.3390/cosmetics8020050 . . . . **35**

Eleni Andreou, Sophia Hatziantoniou, Efstathios Rallis and Vasiliki Kefala
Safety of Tattoos and Permanent Make up (PMU) Colorants
Reprinted from: *Cosmetics* **2021**, *8*, 47, doi:10.3390/cosmetics8020047 . . . . **53**

Laura Kirkbride, Laura Humphries, Paulina Kozielska and Hannah Curtis
Designing a Suitable Stability Protocol in the Face of a Changing Retail Landscape
Reprinted from: *Cosmetics* **2021**, *8*, 64, doi:10.3390/cosmetics8030064 . . . . **63**

Fabian P. Steinmetz, James C. Wakefield and Ray M. Boughton
Fractions of Concern: Challenges and Strategies for the Safety Assessment of Biological Matter in Cosmetics
Reprinted from: *Cosmetics* **2021**, *8*, 34, doi:10.3390/cosmetics8020034 . . . . **71**

About the Editor

Kalliopi Dodou has a First Class Honours BSc in Pharmacy (1999), a PhD in Medicinal Chemistry (2004) and a Postgraduate Diploma in Teaching and Learning in Higher Education. She is a member of the Royal Pharmaceutical Society (MRPharmS), registered with GPharmC, Fellow of the Royal Society of Chemistry (FRSC) and Fellow of the Higher Education Academy (FHEA). She has been an academic at the University of Sunderland since 2004; she started as a lecturer in Pharmaceutics and gradually progressed to senior lecturer (2008) and Reader/Associate Professor (2016) while establishing a new research area for the University on the formulation and quality control of transdermal dosage forms and skin products, initially on the correlation of the rheological behaviour of pressure sensitive adhesives with their adhesive performance on the skin. In 2015 Kalliopi designed the BSc Cosmetic Science course and launched it in 2016; she also designed and launched the MSc Cosmetic Science course in 2020. Kalliopi is the Programme Leader for both courses as well as an elected Member on the Council of the Society of Cosmetic Scientists (SCS) in the UK, a member of the Education Committee and the Scientific Programme Committee of the SCS. Alongside her academic leadership, Kalliopi leads her research group on the design and quality control of skin products and cosmetic formulations; recent projects include the design of novel hydrogel skin patches, the stabilisation of drugs/actives via amorphous solid dispersions, novel quality control testing techniques and sensory testing techniques for the claim substantiation of cosmetic products.

Editorial

Special Issue "Current and Evolving Practices in the Quality Control of Cosmetics"

Kalliopi Dodou

School of Pharmacy, Pharmaceutical and Cosmetic Sciences, University of Sunderland, Sunderland SR1 3SD, UK; kalliopi.dodou@sunderland.ac.uk

Citation: Dodou, K. Special Issue "Current and Evolving Practices in the Quality Control of Cosmetics". *Cosmetics* **2021**, *8*, 100. https://doi.org/10.3390/cosmetics8040100

Received: 8 October 2021
Accepted: 26 October 2021
Published: 29 October 2021

Publisher's Note: MDPI stays neutral with regard to jurisdictional claims in published maps and institutional affiliations.

Copyright: © 2021 by the author. Licensee MDPI, Basel, Switzerland. This article is an open access article distributed under the terms and conditions of the Creative Commons Attribution (CC BY) license (https://creativecommons.org/licenses/by/4.0/).

Quality Control (QC) testing of Cosmetic personal care and fragrance products is a key part of the products' launch to the market. The purpose of QC is to ensure that the product is stable, safe and that its claims are substantiated by scientific data.

Kirkbride et al. [1] critically evaluated current stability QC testing guidelines and techniques based on their industrial experience; they highlighted that the development of reliable stability testing protocols requires a consideration of the product's overall life-cycle and its intended use, concluding that there is a need for product-specific stability strategies.

Barthe et al. [2] provided a comprehensive review of all current in vitro and ex vivo techniques that have replaced the animal studies for the safety QC testing of cosmetic products and cosmetic ingredients. Such techniques include cell culture models, human skin equivalent models and excised human skin. The advantages, challenges and areas for development of these in vitro techniques are discussed in detail, focusing on the safety assessment for genotoxicity, endocrine disruption, dermal absorption, skin and eye irritation.

Steinmetz et al. [3] argued that the ban on animal testing has presented significant challenges in the toxicological safety determination of cosmetic ingredients, especially for those raw materials which are mixtures of plant/botanical extracts with complex chemical compositions. They explain new testing approaches such as the Mode of Action (MoA)-driven testing/analysis and the Threshold of Toxicological Concern (TCC) methodology.

Rakusa and Roskar [4] reported a novel HPCL-UV method for the quantitative QC testing of three actives; Vitamin A, Vitamin E and Coenzyme Q10. The QC testing of commercial anti-ageing products using this novel method revealed labelling discrepancies for these three actives, with actual active concentrations significantly higher or lower than stated. This finding highlighted the need for stricter regulations and quality control testing for active ingredients in cosmetic products.

Andreou et al. [5] elaborated on the safety of tattoo and permanent make up (PMU) colourants. They argue that although there has been a strict quality control of pigment raw materials in recent years, the long-term health risk and toxicological hazards of tattoo inks and PMU colourants need to be further investigated considering that these ingredients are not applied to the skin surface for decorative purposes but are injected into the dermis and reach the systemic circulation.

Biskanaki et al. [6] studied the differences in the expression and quality of skin collagen type -1 (COL I) in healthy, aged, sun exposed, and pathological skin tissues. They observed that sun-exposed skin demonstrates decreased and non-homogeneous COL I expression, which resembles the defective COL 1 expression of benign and cancerous skin lesions. This reinforces the benefits of using skincare products with a sun protection factor.

Claim substantiation testing is unique to cosmetic products. To enable time- and cost-effective quality control testing, assessment methodologies are constantly evolving. In the study conducted by my research group [7] we reported a novel QC method for the determination of refractive indices of creams, using an SPF meter. The RI values then presented the correlation with preliminary skin hydration data after the application of the creams. Such correlations of instrumental data with sensory testing data, can be reliable & cost-effective predictive tools for the cosmetics industry during the initial stages of a product's development.

Institutional Review Board Statement: Not applicable.

Informed Consent Statement: Not applicable.

Conflicts of Interest: The author declare no conflict of interest.

References

1. Kirkbride, L.; Humphries, L.; Kozielska, P.; Curtis, H. Designing a Suitable Stability Protocol in the Face of a Changing Retail Landscape. *Cosmetics* **2021**, *8*, 64. [CrossRef]
2. Barthe, M.; Bavoux, C.; Finot, F.; Mouche, I.; Cuceu-Petrenci, C.; Forreryd, A.; Chérouvrier Hansson, A.; Johansson, H.; Lemkine, G.F.; Thénot, J.-P.; et al. Safety Testing of Cosmetic Products: Overview of Established Methods and New Approach Methodologies (NAMs). *Cosmetics* **2021**, *8*, 50. [CrossRef]
3. Steinmetz, F.P.; Wakefield, J.C.; Boughton, R.M. Fractions of Concern: Challenges and Strategies for the Safety Assessment of Biological Matter in Cosmetics. *Cosmetics* **2021**, *8*, 34. [CrossRef]
4. Temova Rakuša, Ž.; Roškar, R. Quality Control of Vitamins A and E and Coenzyme Q10 in Commercial Anti-Ageing Cosmetic Products. *Cosmetics* **2021**, *8*, 61. [CrossRef]
5. Andreou, E.; Hatziantoniou, S.; Rallis, E.; Kefala, V. Safety of Tattoos and Permanent Make up (PMU) Colorants. *Cosmetics* **2021**, *8*, 47. [CrossRef]
6. Biskanaki, F.; Rallis, E.; Skouras, G.; Stofas, A.; Thymara, E.; Kavantzas, N.; Lazaris, A.C.; Kefala, V. Impact of Solar Ultraviolet Radiation in the Expression of Type I Collagen in the Dermis. *Cosmetics* **2021**, *8*, 46. [CrossRef]
7. Adejokun, D.A.; Dodou, K. A Novel Quality Control Method for the Determination of the Refractive Index of Oil-in-Water Creams and Its Correlation with Skin Hydration. *Cosmetics* **2021**, *8*, 74. [CrossRef]

Article

A Novel Quality Control Method for the Determination of the Refractive Index of Oil-in-Water Creams and Its Correlation with Skin Hydration

Deborah Adefunke Adejokun and Kalliopi Dodou *

School of Pharmacy and Pharmaceutical Sciences, Faculty of Health Sciences and Wellbeing, University of Sunderland, Sunderland SR1 3SD, UK; bg69bo@research.sunderland.ac.uk
* Correspondence: kalliopi.dodou@sunderland.ac.uk; Tel.: +44-(0)191-515-2503

Abstract: The sensory properties of cosmetic products can influence consumers' choice. The accurate correlation of sensory properties, such as skin hydration, with the material properties of the formulation could be desirable. In this study, we aimed to demonstrate a new method for the in vitro measurement of the refractive indices (RIs) of turbid creams. The critical wavelength of each cream was obtained through direct measurement using a sun protection factor (SPF) meter; the wavelength value was then applied in the Sellmeier equation to determine the RI. The results obtained from the in vitro skin hydration measurement for each cream correlated with their RI values. This suggests that RI measurements could be a useful predictive tool for the ranking of creams in terms of their skin hydration effects.

Keywords: sensory testing; refractive index; critical wavelength; turbidity; skin hydration; creams

1. Introduction

Sensory analysis plays a crucial role in the field of cosmetic science. It is used for claim substantiation via subjective users' perception also providing an understanding on how sensory attributes influence consumer's choice and, in turn, the market success of the product [1]. The Organization of Standardization (ISO) allows the properties of a cosmetic product to be described using both qualitative and quantitative methods [2]. This is performed by the selection of a plain descriptive lexicon and a team of well-trained judges to qualify and quantify the test products on the basis of their individual sensory perception via scoring each attribute on a given scale. The statistical evaluation of the collected data can then correlate the scores, assess the overall performance of the product and derive valid claims [3,4]. However, user trials involving scoring scales can be biased due to the subjective nature of the collected data [5,6]. Therefore, there is a need for the development of instrumental techniques that can reliably correlate to and predict sensory properties [7].

In a previous study, we demonstrated how sensory attributes of semisolids, such as pourability, firmness, elasticity, spreadability and stickiness, can be accurately correlated to the rheological measurements of the formulation [7]. In this study, we aimed to develop further such correlations.

The refractive index (RI) of a material is a measure of the velocity of light in vacuum divided by the velocity of light crossing the material. The RI has a wide range of applications, from the estimation of drug concentration present in a sample to the opaqueness or turbidity of the sample [8]. Changes in the RI of the skin have been shown to correlate to skin hydration after the application of a moisturizer; this is because hydration renders the skin less opaque, i.e., more translucent, resulting in a decrease in the RI [9–11]. The measurements of the RIs of cosmetic creams could potentially be correlated to their short-term skin hydration effects, assuming that less turbid creams with a low RI will

have a high water content or/and dissolved actives with a hydrating effect. Because of the turbid nature of creams, a traditional UV spectrophotometer cannot be used to measure their RIs [12,13], and the optical coherence tomography technique cannot elaborate their RIs on a routine basis [9]; therefore, there is a need for developing a simple method to accurately and routinely measure the RIs of creams and turbid samples.

The objectives of the present study were as following: (i) to develop a method for the direct and accurate measurement of the RIs of turbid formulations; and (ii) to investigate if there is a correlation between the RIs of creams and their skin hydration effects.

2. Materials and Methods

2.1. Materials

The active ingredient (X), cholesterol, span65 and solutol HS-15 were obtained from Sigma-Aldrich, Inc. (Gillingham, UK). Baobab oil was purchased from Aromatic Natural Skin Care (Forres, UK), and jojoba and coconut oil were bought from SouthernCross Botanicals (Knockrow, Australia). The Emulsifying Wax was obtained from CRODA International Plc (Goole, East Yorkshire, UK). Other excipients of the cream and Tris buffer solutions were of analytical grade.

2.2. Methods

2.2.1. Preparation of the Oil-in-Water Creams

Oil-in-water creams IA–IVA and IB–IVB used in this study were formulated as explained in our previous paper [7].

Four active-containing oil-in-water creams (labelled A) and their controls (labelled B; without active ingredients) were prepared. Each cream contained the following oil combinations: I (volume ratio of 8% jojoba and baobab oils, 1:1)—water phase (85%), oil phase (10%) and emulsifier (5%); II (volume ratio of 10% jojoba and baobab oil, 1:1), III (volume ratio of 10% jojoba oil and coconut oil, 1:1) and IV (volume ratio of 10% baobab and coconut oil, 1:1)—water phase (83%), oil phase (12%) and emulsifier (5%). The composition of each cream is shown in Table 1.

Table 1. Ingredients and % (w/w) composition of each cream formulation.

Cream Phases	Ingredients	Composition of Each Cream Formulation (% w/w)							
		IA	IB	IIA	IIB	IIIA	IIIB	IVA	IVB
Oil phase	Stearyl alcohol					1	1	1	1
	Jojoba oil	4	4	5	5	5	5		
	Baobab oil	4	4	5	5			5	5
	Coconut oil					5	5	5	5
Water phase	Glycerine	5	5			5	5	5	5
	Propylene glycol			5	5				
	Water	73.7	78.7	71.7	76.7	71.7	76.7	71.7	76.7
Active	Entrapped active ingredient	5		5		5		5	

2.2.2. Measurement of Refractive Index Using an SPF Analyser

A sun protection factor (SPF) analyser, SPF-290AS (SolarLight®, Glenside, PA, USA), was used to obtain the individual transmittance wavelength of each cream sample. The RI of each cream was then calculated using the Sellmeier equation (Equation (1)):

$$n^2(\lambda) = 1 + (B_1\lambda^2/\lambda^2 - C_1) + (B_2\lambda^2/\lambda^2 - C_2) + (B_3\lambda^2/\lambda^2 - C_3), \qquad (1)$$

where

n is the refractive index (RI),
λ is the wavelength of the test sample Determined by a SPF meter.

The coefficients of the Sellmeier equation for a fused silica/silicon substrate can be shown as following [14]:

$B_1 = 0.696166300$; $C_1 = 4.67914826 \times 10^{-3} \ \mu m^2$;
$B_2 = 0.407942600$; $C_2 = 1.35120631 \times 10^{-2} \ \mu m^2$;
$B_3 = 0.897479400$; $C_3 = 97.9340025 \ \mu m^2$.

A wavelength measurement on the SPF working at the wavelength of 290 m was carried out by initially placing an empty silicon substrate or a transpore tape (to imitate the skin surface) in the optical path to acquire a reference scan. The substrate was then loaded with the test sample at 2.0 µL/cm², spread out in a unidirectional motion, allowed to dry for 15 min and returned to the optical path. Six different scans were taken by a monochromator over the wavelength region of 380–500 nm, and an average scan was produced. The SPF software factors out the reference scan data, resulting in the transmittance of only the measured sample. This experiment was conducted in triplicate for each cream at room temperature.

2.2.3. Skin Hydration

A preliminary self-evaluation of the hydrating effects of the creams was carried out by the researcher using a skin hydration meter (Moisture meter SC Compact, Delfin, UK). Hands were washed with a soap and dried for 5 min. The moisture meter was placed on the back of the hand and held in a steady position, until a measurement was taken at T0 (t = 0) and the value was recorded. A small amount of cream was then applied in a circular motion to the same location of the original reading, and after 5 min, a second measurement was taken at T5 (t = 5). This process was repeated in triplicate for each cream.

2.2.4. Statistical Analysis

A statistical evaluation of results was carried out using the IBM SPSS software. To indicate whether any significant difference ($p < 0.05$) existed in the skin hydration after the cream application, a paired sample t-test was used to compare the results before and after measurements.

3. Results and Discussion

Correlation between RI Measurements and Skin Hydration Results

Table 2 shows the calculated RI values for all cream models. All creams were opaque or turbid with RI values greater than the RI of full fat milk cream, which is 1.38810 [12]. The least turbid creams were models IA and IB, having the lowest RI values (Table 2).

Table 2. The critical wavelengths of all models taken in triplicate and their refractive index (RI) values.

Model	Wavelength 1	Wavelength 2	Wavelength 3	Mean Wavelength (nm)	±SD	Wavelength (µm)	RI Value
IA	387.7	387.8	387.8	387.8	0.06	0.3878	2.12377
IB	387.7	387.7	387.8	387.7	0.06	0.3877	2.12378
IIA	385.0	384.9	385.0	385.0	0.06	0.3850	2.12397
IIB	385.1	385.0	385.0	385.0	0.06	0.3850	2.12397
IIIA	385.4	385.4	385.4	385.4	0	0.3854	2.12393
IIIB	385.5	385.6	385.5	385.5	0.06	0.3855	2.12393
IVA	385.9	385.5	385.9	385.8	0.23	0.3858	2.12391
IVB	385.9	385.9	385.9	385.9	0	0.3859	2.12392

Table 3 shows the % increase in skin hydration after the application of each cream. The paired sample *t*-test revealed there was a statistically significant difference in skin hydration (*p*-value < 0.05) before and after the measurements of skin hydration for most cream models.

Table 3. Measurement values of skin hydration in percentage (%) before and after the measurements of skin hydration (T_0 vs. T_5).

Cream Model/ Number	1		2		3		Mean		*p*-Value	% Increase in Hydration
	Before (T_0)	After (T_5)	Before (T_0)	After (T_5)	Before (T_0)	After (T_5)	Before (T_0/±SD)	After (T_5/±SD)		
IA/1	51.2	59.4	29.6	55.9	55.5	59.8	45.4/±13.9	58.4/±2.1	0.09	28.63
IB/2	36.4	52.3	59.8	67.2	51.7	65.3	49.3/±11.9	61.6/±8.1	0.02	24.94
IIA/3	58.7	62.4	58.0	67.4	51.5	55.6	56.1/±4.0	61.8/±5.9	0.04	10.16
IIB/4	44.5	57.1	66.8	67.2	48.0	58.7	53.1/±11.9	61.0/±5.4	0.08	14.87
IIIA/5	67.4	72.0	60.3	68.6	54.9	63.8	60.9/±6.3	68.1/±4.1	0.01	11.82
IIIB/6	65.2	65.7	70.5	77.1	62.1	73.0	65.9/±4.2	71.9/±5.7	0.09	9.10
IVA/7	58.2	64.7	55.6	60.8	65.9	70.5	59.9/±5.4	65.3/±4.9	0.005	9.01
IVB/8	61.0	74.1	66.4	73.9	59.4	66.5	62.3/±3.7	71.5/±4.3	0.02	14.76

According to these preliminary results, creams IA and IB showed the highest increase in hydration alongside the lowest RI values, therefore confirming the hypothesis that creams with a lower RI have a higher hydrating effect. This correlation is depicted in Figure 1.

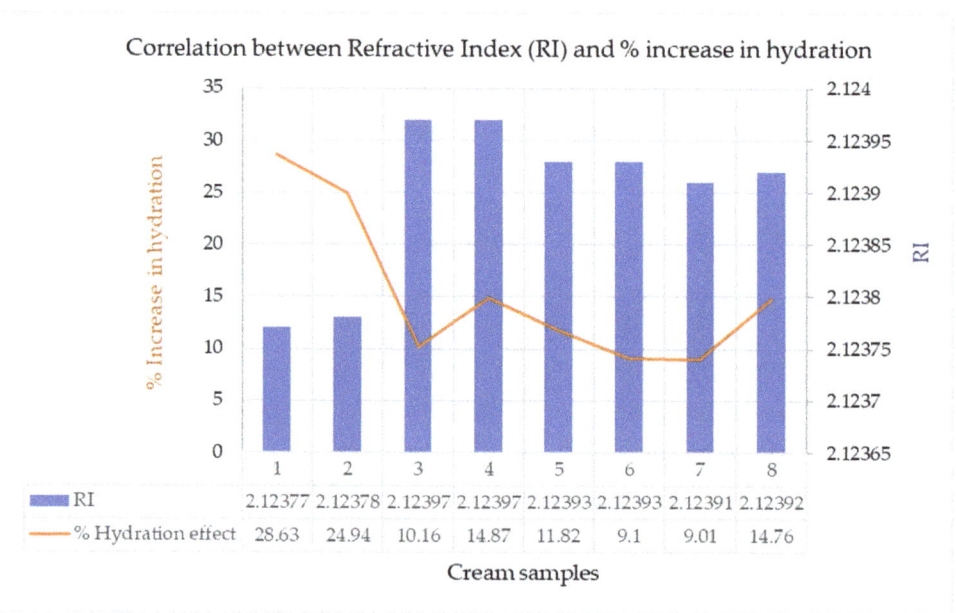

Figure 1. Relationships between RI and % increase in hydration. Cream samples (IA to IVB) are shown in numerical order of 1–8.

The high skin hydration effect seen in models IA and IB could be attributed to the higher water contents in their compositions compared to the rest of the creams (Table 1). This effect can also be attributed to the humectant used and its compatibility with jojoba and baobab oil [15], considering that cream models IIA and IIB contained the same type of oils as IA and IB but with a different humectant. In fact, the turbid nature and low hydration effect of models IIA and IIB could be the effect of propylene glycol (Table 1) in comparison to the glycerin present in IA and IB. Based on these observations, the water content and the type of humectant seemed to be the determining factors in the hydration effect. Changes in these parameters can be detected by RI measurements, via changes in ingredients' solubility/compatibility, therefore explaining the observed correlation between RI/turbidity and hydration effects. Similarly, the high RI and the low hydration of models III and IV can be attributed to the presence of stearyl alcohol in models III and IV (Tables 1–3) and its incompatibility with glycerine and/or the oils in these formulas.

4. Conclusions

In this study, we reported our preliminary findings on a new method for the prediction of the skin hydration effects of creams by measuring their RIs, using the Sun Protection Factor-290 Automated System (SPF-290AS). This newly developed method using the SPF equipment and the Sellmeier equation is highly sensitive and simple and allows the determination of the RIs of turbid samples without dilution. The correlation of the RIs of the creams with their skin hydration effects could be a predictive tool for the cosmetics industry to provide useful information on the hydration effect of creams before embarking on user trials and/or confirming the results from user trials. Further studies should include a large-scale user trial to validate this predictive tool, also including different types of semisolid formulations (water-in-oil and gels) to explore the extent of its applicability.

Author Contributions: Conceptualization, D.A.A. and K.D.; methodology, D.A.A.; validation, D.A.A. and K.D.; formal analysis, D.A.A. and K.D.; investigation, D.A.A.; resources, K.D.; data curation, D.A.A.; writing—original draft preparation, D.A.A.; writing—review and editing, K.D.; supervision, K.D.; project administration, K.D. All authors have read and agreed to the published version of the manuscript.

Funding: This research received no external funding.

Institutional Review Board Statement: The study was conducted according to the guidelines of the Declaration of Helsinki, and approved by the Institutional Review Board (or Ethics Committee) of The University of Sunderland (protocol code 006180 and date of approval: 20 February 2020).

Informed Consent Statement: Informed consent was obtained from all subjects involved in the study.

Data Availability Statement: Data is contained within the article.

Conflicts of Interest: The authors declare no conflict of interest.

References

1. Lawless, H.T. *Quantitative Sensory Analysis: Psychophysics, Models and Intelligent Design*; John Wiley & Sons: Hoboken, NJ, USA, 2013; p. 12.
2. ISO—ISO 13299:2016—Sensory Analysis—Methodology—General Guidance For Establishing A Sensory Profile. Available online: https://www.iso.org/standard/58042.html (accessed on 20 February 2020).
3. Murray, J.M.; Delahunty, C.M.; Baxter, I.A. Descriptive sensory analysis: Past, present and future. *Food Res. Int.* **2001**, *34*, 461–471. [CrossRef]
4. Pensé-Lhéritier, A.M. Recent developments in the sensorial assessment of cosmetic products: A review. *Int. J. Cosmet. Sci.* **2015**, *37*, 465–473. [CrossRef] [PubMed]
5. Montenegro, L.; Rapisarda, L.; Ministeri, C.; Puglisi, G. Effects of lipids and emulsifiers on the physicochemical and sensory properties of cosmetic emulsions containing vitamin E. *Cosmetics* **2015**, *2*, 35–47. [CrossRef]
6. Varela, P.; Ares, G. Sensory profiling, the blurred line between sensory and consumer science. A review of novel methods for product characterization. *Food Res. Int.* **2012**, *48*, 893–908.
7. Adejokun, D.A.; Dodou, K. Quantitative Sensory Interpretation of Rheological Parameters of a Cream Formulation. *Cosmetics* **2020**, *7*, 2. [CrossRef]
8. Singh, S. Refractive index measurement and its applications. *Phys. Scr.* **2002**, *65*, 167. [CrossRef]
9. Sand, M.; Gambichler, T.; Moussa, G.; Bechara, F.G.; Sand, D.; Altmeyer, P.; Hoffmann, K. Evaluation of the epidermal refractive index measured by optical coherence tomography. *Skin Res. Technol.* **2006**, *12*, 114–118. [CrossRef] [PubMed]

10. Chadwick, A.C.; Kentridge, R.W. The perception of gloss: A review. *Vis. Res.* **2015**, *109*, 221–235. [CrossRef] [PubMed]
11. Ezerskaia, A.; Ras, A.; Bloemen, P.; Pereira, S.F.; Urbach, H.P.; Varghese, B. High sensitivity optical measurement of skin gloss. *Biomed. Opt. Express* **2017**, *8*, 3981–3992. [CrossRef] [PubMed]
12. Calhoun, W.R.; Maeta, H.; Roy, S.; Bali, L.M.; Bali, S. Sensitive real-time measurement of the refractive index and attenuation coefficient of milk and milk-cream mixtures. *J. Dairy Sci.* **2010**, *93*, 3497–3504. [CrossRef] [PubMed]
13. Walstra, P. *Dairy Technology: Principles of Milk Properties and Processes*; CRC Press: Boca Raton, FL, USA, 1999; p. 23.
14. Sellmeier Equation—Wikipedia. Available online: https://en.wikipedia.org/wiki/Sellmeier_equation (accessed on 20 February 2020).
15. Sagiv, A.E.; Dikstein, S.; Ingber, A. The efficiency of humectants as skin moisturizers in the presence of oil. *Skin Res. Technol.* **2001**, *7*, 32–35. [CrossRef] [PubMed]

Article

Quality Control of Vitamins A and E and Coenzyme Q10 in Commercial Anti-Ageing Cosmetic Products

Žane Temova Rakuša and Robert Roškar *

Faculty of Pharmacy, University of Ljubljana, Aškerčeva cesta 7, 1000 Ljubljana, Slovenia; zane.temova.rakusa@ffa.uni-lj.si
* Correspondence: robert.roskar@ffa.uni-lj.si; Tel.: +386-1-4769-500

Citation: Temova Rakuša, Ž.; Roškar, R. Quality Control of Vitamins A and E and Coenzyme Q10 in Commercial Anti-Ageing Cosmetic Products. *Cosmetics* **2021**, *8*, 61. https://doi.org/10.3390/cosmetics8030061

Academic Editor: Kalliopi Dodou

Received: 14 May 2021
Accepted: 21 June 2021
Published: 25 June 2021

Publisher's Note: MDPI stays neutral with regard to jurisdictional claims in published maps and institutional affiliations.

Copyright: © 2021 by the authors. Licensee MDPI, Basel, Switzerland. This article is an open access article distributed under the terms and conditions of the Creative Commons Attribution (CC BY) license (https://creativecommons.org/licenses/by/4.0/).

Abstract: Vitamins A and E and coenzyme Q10 are common ingredients in anti-ageing cosmetic products. Within this study, we evaluated the quality of commercial cosmetics with vitamin A (35 products), vitamin E (49 products), and coenzyme Q10 (27 products) by using validated HPLC–UV methods. Vitamin A was determined as retinol, retinyl palmitate, retinyl propionate, β carotene, and hydroxypinacolone retinoate in concentrations ranging from 950 ng/g to 19 mg/g. Total vitamin A contents, expressed with retinol equivalents, ranged from 160 ng/g to 19 mg/g, and were above the maximum concentration recommended by the SCCS in six of the 35 tested cosmetics. The content-related quality control of 10 cosmetics with specified vitamin A content revealed significant deviations (between 0% and 400%) of the label claim. Vitamin E was determined as both tocopherol and tocopheryl acetate in concentrations between 8.5 µg/g and 16 mg/g. Coenzyme Q10 was determined as ubiquinone in 24 tested cosmetics, which labelled it, in concentrations between 4.2 µg/g and 100 µg/g. Labelling irregularities were observed in all three active compound groups, resulting in a significant share (42%) of improperly labelled cosmetic products. The results of this study reveal the need for stricter cosmetics regulation and highlight the importance of their quality control, especially by evaluating the contents of the active compounds, in their efficacy and safety assurance.

Keywords: active compounds; assay; cosmeceutics; functional cosmetics; HPLC–UV; labelling; retinoids; tocopherol; ubiquinone; β carotene

1. Introduction

The topical application of fat-soluble vitamins A and E and coenzyme Q10 has various beneficial effects on the skin. Therefore, these three groups are important ingredients in the cosmetic industry [1,2]. The widespread use of vitamin E over the past several decades is mostly associated with its antioxidant activity [3]. Vitamin E is used in cosmetics as a cosmetically active ingredient (occlusive, humectant, emollient, and miscellaneous agent) [4] or as a stabilizer of other, unstable components of the cosmetic product [5,6]. Because of its antioxidant activity, topically applied vitamin E is effective in the treatment of skin conditions and diseases caused by oxidative stress, including UV-induced erythema and edema, sunburns, and lipid peroxidation [1,2]. It is also an effective anti-ageing agent [7,8]. Vitamin E is most commonly found in cosmetics in its active form, α-tocopherol, or more stable esterified form, tocopheryl acetate, which requires hydrolysis to the active form upon skin penetration [4]. Despite differing data on the extent of this conversion in the skin, most studies disclose the higher antioxidant activity of α-tocopherol compared to its esters [4,9–12]. Vitamin E may be found in a wide range of concentrations, from 0.0001% to 36% in cosmetic products on the market [13].

Retinoids are effective in the topical treatment of acne, hyperpigmentation, psoriasis, and skin-aging, and are therefore active ingredients in a variety of cosmetic products, especially as anti-ageing agents. The most common vitamin A forms found in cosmetics include retinol and its esters, retinyl palmitate and acetate, as well as β carotene. Analogously to vitamin E esters, vitamin A esters also require hydrolytic conversion to retinol, which is further metabolized to retinal and then to the active form—retinoic acid. Therefore, retinoid activity after topical application depends on the metabolic closeness to the active form and decreases in the following order: retinoic acid > retinal > retinol > retinyl esters [14]. Due to the possible risk of teratogenicity, retinoic acid is banned in cosmetic products in the EU [15]. Despite their poor activity, retinyl esters, especially retinyl palmitate, are commonly used in cosmetics due to their stability [14,16]. Due to safety reasons, the Scientific Committee on Consumer Safety, Secretariat at the European Commission, Directorate General for Health and Food Safety recommends a maximum retinoid concentration of 0.05% retinol equivalents (RE) in body lotions and 0.3% RE in hand and face creams, as well as other leave-on or rinse-off products for cosmetics in the EU [14]. However, cosmetics with significantly higher retinoid contents are found on the EU market.

Coenzyme Q10 is an endogenous nonvitamin lipophilic antioxidant, which is often analytically evaluated alongside fat-soluble vitamins, due to its lipophilic structure and activities in the human body [17]. Coenzyme Q10 is also an important antioxidant in the skin [1,18]. However, its skin levels decline with age and exposure to UV irradiation [19]. Topical coenzyme Q10 application is effective in the replenishment of its skin levels and thus provides skin protection and prevents skin inflammation, UV-induced erythema, and skin cancer [18,20,21]. Coenzyme Q10, in its ubiquinone form, is a popular ingredient in anti-ageing cosmetics, in which it is usually found in concentrations $\leq$ of 0.05% [22].

The efficacy of cosmetic products is directly associated with their quality. As discussed above, the efficacy depends on the form of the active ingredient (e.g., vitamin A or E esters), and also on their content, which is generally low (<1%). Another important challenge is the instability of these compounds, causing possible losses during manufacture and storage, leading to their even lower contents or complete loss [8,23,24]. Therefore, a prerequisite for their quality control is appropriate analytical methodology. Several analytical methods for the determination of a single retinoid [25–30] or retinoids in different forms [31–33] in topical formulations may be found in the literature, including two methods [34,35] for the quality control of specific vitamin A forms commonly found in cosmetics. The simultaneous determination of coenzyme Q10 and vitamin E (mostly in the form of tocopheryl acetate) in pharmaceutical products has been described in the literature [17,36], but to our knowledge has not been applied in the field of cosmetics. Within this study, we aimed to evaluate the quality of a significant number of commercial anti-ageing leave-on cosmetic products with vitamin A and E and coenzyme Q10 by applying appropriate analytical methodologies, including a novel method for the quality control of coenzyme Q10 and vitamin E, as tocopherol or tocopheryl acetate. We approached their quality control following the principles of the quality control of pharmaceuticals—by evaluation of the accuracy of their labelling, content determination, and comparison to the quantitative label claims in some cosmetics.

2. Materials and Methods
2.1. Chemicals and Reagents

The following vitamins were obtained from Sigma-Aldrich (Steinheim, Germany): all-*trans*-retinol (R) ($\geq$99%), all-*trans*-retinyl palmitate (R-palm) ($\geq$99%), β carotene (β-car) ($\geq$99%), ($\pm$)-α-tocopherol (E) ($\geq$96%), and DL-α-tocopherol acetate (E-ac) ($\geq$96%). Retinyl acetate (R-ac) ($\geq$97%) and coenzyme Q10 as ubiquinone (Q10) ($\geq$99%) were purchased from Carbosynth (Berkshire, UK). Butylated hydroxytoluene (BHT) and HPLC-grade acetonitrile (ACN), tetrahydrofuran (THF), and *n*-hexane were obtained from Sigma-Aldrich (Steinheim, Germany). Ultra-pure water (MQ) was obtained through a Milli-Q water purification system A10 Advantage (Millipore Corporation, Bedford, MA, USA).

2.2. Instrumentation and Chromatographic Conditions

An Agilent 1100/1200 series HPLC system (Agilent Technologies, Santa Clara, CA, USA) equipped with a UV–VIS detector and ChemStation data acquisition system was utilized. The analysis was performed on a reversed-phase Luna C18 (2) 150 mm × 4.6 mm, 3 µm particle size column (Phenomenex, Torrance, CA, USA) at 40 °C using MQ (mobile phase A), ACN (mobile phase B), and ACN:THF (50:50, v/v) (mobile phase C) at a flow-rate of 1 mL/min.

Vitamin E (tocopherol and tocopheryl acetate) and coenzyme Q10 (ubiquinone) were analyzed using a gradient elution with the following gradient for their chromatographic separation: (time (min); % A; % B; % C): (0; 10; 10; 80), (5.5; 10; 10; 80), (7.0; 3; 5; 92), (10.0; 3; 5; 92), (10.1; 10; 10; 80). The detection wavelength was 280 nm. The injection volume was adjusted to the content of the examined analytes in the samples and was between 10 µL and 20 µL.

The examined retinoids (retinol, retinyl palmitate, retinyl acetate, β carotene, hydroxypinacolone retinoate, and retinyl propionate) were analyzed according to a validated method [34] using the following gradient program: (time (min); % A; % B; % C): (0; 10; 5; 85), (4; 5; 5; 90), (8; 5; 5; 90), (8.1; 10; 5; 85). Detection was carried out at 325 nm for retinol, retinyl acetate, retinyl palmitate, hydroxypinacolone retinoate, and retinyl propionate, and at 450 nm for β carotene. Injection volume was adjusted to the amount of retinoids in the tested products and was between 5 µL and 40 µL.

2.3. Preparation of Standard Solutions

Retinol, retinyl acetate, retinyl palmitate, β carotene, tocopherol, tocopheryl acetate, and coenzyme Q10 stock standard solutions were prepared fresh daily by dissolving appropriate amounts of the individual standard in a mixture of ACN and THF (50:50, v/v) in the case of vitamin E and coenzyme Q10, and n-hexane containing 500 mg/L BHT in the case of retinoids. Calibration standards and quality control (QC) solutions (in triplicate) were prepared by dilution of the individual stock standard solutions with the same solvent (Table 1). The retinoid solutions were evaporated to dryness under a stream of nitrogen at 40 °C (TurboVap LV, Caliper, Hopkinton, MA, USA) and reconstituted with a mixture of ACN and THF (50:50, v/v) with 150 mg/L BHT to obtain calibration standards and QC solutions as presented in Table 1. Standard solutions with lower concentrations than those presented in Table 1 were also prepared for confirmation of the limit of determination (LOD) and limit of quantification (LOQ). The prepared standard solutions were immediately analyzed.

Table 1. Concentrations (mg/L) of calibration standards and QC solutions.

	R	R-palm	R-ac	β-car	E	E-ac	Q10
Calibration standards	0.25	0.38	0.25	0.25	8.00	8.00	2.50
	1.00	1.50	1.00	1.00	40.0	40.0	12.5
	10.0	15.0	10.0	10.0	80.0	80.0	25.0
	25.0	37.5	25.0	25.0	320	320	100
	75.0	113	75.0	75.0	480	480	150
	100	150	100	100	800	800	250
QC solutions	5.00	7.50	5.00	5.00	16.0	16.0	5.00
	15.0	22.5	15.0	15.0	160	160	50.0
	50.0	75.0	50.0	50.0	640	640	200

2.4. Method Validation

Both utilized HPLC–UV methods were validated following the ICH guidelines Q2(R1) [37] in terms of specificity, linearity, precision, accuracy, LOD, LOQ, sample stability, sample preparation repeatability, and recovery. Specificity was assessed in both standard solutions (individual standards, their mixtures, all used solvents, some common ingredients in cosmetics-purified water, white petroleum jelly, liquid paraffin, cetyl and stearyl alcohol, macrogol cetostearyl ether, and benzyl alcohol) and in chromatograms of the tested cosmetic products, which were evaluated for interferences.

Linearity was assessed by a linear regression model of the individual analyte calibration standards, which were prepared and analyzed during three consecutive days (Table 1). The acceptance criterion was coefficient of determination (R^2) > 0.999. The injection volumes during validation were 10 µL for retinol, retinyl acetate, and retinyl palmitate, and 20 µL for β carotene, tocopherol, tocopheryl acetate, and coenzyme Q10.

Accuracy and precision were evaluated intra- and inter-day on three QC levels, during three consecutive days (Table 1). Accuracy was determined as a ratio between the concentration calculated from the regression line and the actual concentration and was set to 100 ± 10%. Precision was determined as a relative standard deviation (RSD) of the three QC solutions on each concentration level and was set at ≤5%. Injection repeatability was determined by six consecutive injections of the medium QC solution and was set to ≤2%.

The LOD and LOQ were determined by a signal-to-noise ratio of 3:1 and 10:1, respectively, and were evaluated in chromatograms of standard solutions with known low concentrations and blank samples. Both values were confirmed by the analysis of standard solutions with comparable concentrations. LOD and LOQ values are provided as ng of the analyte per one gram of cosmetic product and were calculated for the most concentrated samples according to the sample preparation procedures (see Section 2.6. Analysis of the commercial cosmetic products).

The stability of the evaluated analytes was assessed in QC solutions on all three levels, which were stored at 8 °C for up to 24 h. Sample stability was calculated as a share of the initial response and was expected to be within 100 ± 5%.

Sample preparation repeatability was assessed by preparation of all tested products in triplicate and calculating the RSD between them, which was set at ≤5%.

Method recovery was assessed by the addition of the evaluated analyte to a cosmetic product with its significant amounts and separate analysis of the cosmetic product without addition and of the standard solution containing the added analyte amount in the extraction solvent. All samples were prepared in triplicate. Average recoveries were calculated by the following equation: recovery (%) = 100 × (concentration found in spiked sample−concentration found in the non-spiked sample)/added concentration. They were set at 100 ± 10%.

2.5. Selection and Overview of the Analyzed Commercial Cosmetic Products

Within this study, we evaluated anti-ageing leave-on facial cosmetic products, containing vitamin A, E, and/or coenzyme Q10. The cosmetic products were purchased between 2015 and 2021. All products were obtained locally on the Slovenian market, including grocery stores, drug stores, pharmacies, and on the Internet. To provide representative samples, products in various formulations (day and night creams, serums, eye creams, anti-ageing concentrates, and face tonics) and labelled with different forms of vitamin A and E were correspondingly included. One of the selection criteria was also the quantitative declaration of the content of the evaluated active ingredients on the cosmetic products. The obtained cosmetic products were categorized into five price ranges, considering their retail price in Slovenia, calculated to a uniform volume of 50 mL. An overview of the tested products, indicating the labelled forms of vitamins A and E and coenzyme Q10, as well as their forms and the price ranges, are provided in Table 2. Five cosmetic products (5, 6, 23, 40, and 54 in Table 2) have been previously analyzed [34].

Table 2. Overview of the tested cosmetic products—their form, price range, and labelled vitamin A, vitamin E, and coenzyme Q10.

No.	Form	Vitamin A	Vitamin E	Coenzyme Q10	Price (€/50 mL)
1	C	0.0055% R-palm			≤5
2	DC	β-car	E	ubiquinone	≤5
3	NC	β-car	E	ubiquinone	≤5
4	DC	R-palm	E	ubiquinone	≤5
5	C	R-palm	E	ubiquinone	≤5
6	C	R-palm	E	ubiquinone	≤5
7	C	R-palm			≤5
8	C	R-palm	E		≤5
9	NC	R	E-ac		≤5
10	DC	R, R-palm			≤5
11	DC		E-ac	ubiquinone	≤5
12	DC		E-ac	ubiquinone	≤5
13	S		E-ac	ubiquinone	≤5
14	NC		E, E-ac	ubiquinone	≤5
15	DC		E-ac	ubiquinone	≤5
16	S		E-ac	ubiquinone	≤5
17	NC	β-car	E	ubiquinone	≤5
18	C	R-palm	E, E-ac	ubiquinone	≤5
19	C	R-palm	E, E-ac		≤5
20	DC		E, E-ac	ubiquinone	≤5
21	C		E, E-ac	ubiquinone	≤5
22	T	R-palm			≤5
23	S	1% R			5–15
24	S	0.2% R			5–15
25	DC	R-palm	E-ac		5–15
26	C	R-palm, β-car	E, E-ac		5–15
27	EC			ubiquinone	5–15
28	AC		E-ac	ubiquinone	5–15
29	DC	R-palm	E, E-ac	ubiquinone	5–15
30	S	R, R-palm	E, E-ac		5–15
31	S		E-ac	ubiquinone	5–15
32	C	R-palm	E	ubiquinone	5–15
33	DC			ubiquinone	5–15
34	NC		E-ac	ubiquinone	5–15
35	C			ubiquinone	5–15
36	C			ubiquinone	5–15
37	C		E-ac		5–15
38	S	R-palm	E, E-ac	ubiquinone	5–15
39	S	0.5% R			15–30
40	C	2% HRP, R			15–30
41	C		E, E-ac		15–30
42	AC	R-palm	E, E-ac		15–30
43	DC	β-car	E, E-ac	ubiquinone	15–30
44	NC			ubiquinone	15–30
45	C		E, E-ac	ubiquinone	15–30
46	C		E		15–30
47	C		E		15–30
48	S	R-palm	E, E-ac		15–30
49	DC		E		15–30
50	C	β-car	E, E-ac		15–30
51	C	β-car	E, E-ac		15–30
52	C		E, E-ac		15–30
53	C	0.5% R, R-prop	E, E-ac		30–60
54	C	0.2% R			30–60
55	AC			ubiquinone	30–60

Table 2. *Cont.*

56	C	R-palm	E-ac	30–60
57	C		E	30–60
58	S		E	30–60
59	EC		E	30–60
60	NC		E	30–60
61	S		E	30–60
62	AC	R, R-palm	E	30–60
63	S		E	30–60
64	S		E	30–60
65	DC		E-ac	30–60
66	S	0.03% R, R-palm		60–125
67	C	1% R		60–125
68	S	2.5% R		60–125
69	C		E-ac	60–125
70	C	R, R-palm	E	60–125
71	EC		ubiquinone	60–125
72	C		ubiquinone	60–125
73	C		E, E-ac	60–125

AC—anti-ageing concentrate; C—cream; DC—day cream; E—tocopherol; E-ac—tocopheryl acetate; EC—eye cream; HRP—hydroxypinacolone retinoate; NC—night cream; R—retinol; R-palm—retinyl palmitate; R-prop—retinyl propionate; S—serum; T—face tonic; β-car—β carotene.

2.6. Analysis of the Commercial Cosmetic Products

All tested cosmetic products were analyzed within their shelf-life, immediately after opening, in triplicate. Due to a time-lapse between the establishment of both analytical methods and the time of analysis, vitamin A, E, and coenzyme Q10 were not evaluated in 2 of the 37 cosmetics, 5 of the 53 cosmetics, and 4 of the 31 cosmetics, respectively, which labelled their presence. To accurately evaluate the content of vitamins A and E and coenzyme Q10, preliminary testing was initially performed, based on which the sample preparation procedure was adjusted to the individual cosmetic product.

2.6.1. Sample Preparation for the Analysis of Vitamin E and Coenzyme Q10

Samples for the analysis of tocopherol, tocopheryl acetate, and coenzyme Q10 were prepared by weighing a certain amount (between 200 and 1000 mg of the cosmetic product) into a plastic tube. A predefined amount (2, 5, or 10 mL) of a mixture of ACN and THF (75:25, v/v %) was added to the cosmetic product, followed by vortex mixing (3 min), sonication (15 min), additional vortex mixing (2 min), and centrifuging (4130× g, 25 °C, 10 min). The samples were filtered through a 0.45 μm Minisart® RC filter (Sartorious, Göttingen, Germany) before analysis.

2.6.2. Sample Preparation for the Analysis of Vitamin A

Samples for evaluation of vitamin A content were prepared according to a validated procedure for their analysis and quantification [34]. Samples from the tested semi-solid cosmetic products were prepared by initial weighing of the cosmetic product (75–1000 mg) into a plastic tube. Acetonitrile (2 mL) was added to the samples, followed by their sonication (5 min). Then, n-hexane (8 mL) with 500 mg/L BHT was added to the samples, which were further vortexed (5 min) and centrifuged (4130× g, 25 °C, 10 min). Part of the supernatant (0.5 mL–2.0 mL) was evaporated to dryness under a stream of nitrogen at 40 °C (TurboVap LV, Caliper, Tokyo, Japan). Dry residues were reconstituted with a mixture of ACN and THF (50:50, v/v %) with 150 mg/L BHT (0.5 mL–2.0 mL), sonicated (10 min), and vortexed (1 min). If needed, the samples were centrifuged (16 200× g, 25 °C, 5 min) before analysis.

Samples of the tested liquid cosmetic products were prepared by their dilution by 5- to 500-fold with a mixture of ACN and THF (50:50, v/v %) with 150 mg/L BHT. The samples were homogenized by sonication (10 min) and vortex mixing (5 min). If needed, the samples were centrifuged (16,200× g, 25 °C, 5 min) before analysis.

2.6.3. Quantification of Vitamins A and E and Coenzyme Q10

The contents of vitamins A and E and coenzyme Q10 in the tested cosmetic products were calculated from their linear regression lines. Due to the lack of hydroxypinacolone retinoate and retinyl propionate standards, their content was assessed based on retinyl acetate, which is structurally their most similar retinoid based on previous confirmation of their structural identity by LC-MS.

The analytically determined contents are presented as an average (AV) ± standard error of the mean (SEM), n = 3. Vitamin E contents are provided as a mass percentage (% m/m). The contents of vitamin A and coenzyme Q10 are provided as a mass percentile (‰ m/m), due to their lower contents. The tested cosmetic products were numbered consecutively within the specific categories (e.g., cosmetic products with vitamin E). The numbers in Table 2 and Figures 3–7 are not correlated between different figures and Table 2, and do not represent identification numbers for the individual products.

3. Results

3.1. Validation of the HPLC–UV Methods

The utilized HPLC–UV methods were validated following the ICH guidelines Q2(R1) in terms of specificity, linearity, precision, accuracy, LOD, LOQ, sample stability, sample preparation repeatability, and recovery. The specificity of the method was confirmed for all evaluated analytes as no interferences derived from the used solvents and the evaluated reagents, other evaluated analytes, or the cosmetic products were detected at their retention times and detection wavelength. A representative chromatogram of a standard mixture of retinol, retinyl acetate, and retinyl palmitate at 325 nm, β carotene standard solution at 450 nm, and standard mixture of tocopherol, tocopheryl acetate, and coenzyme Q10 at 280 nm, as well as some of the analyzed cosmetic products are provided in the Supplementary Materials (Figures S1–S7). Linearity was confirmed over the evaluated concentration ranges (Table 3). The methods' LOD and LOQ were determined based on the signal-to-noise ratio. The methods were found sufficiently sensitive for the determination of the evaluated analytes in cosmetic products (Table 3). The sensitivity may be additionally increased by adjustments in the sample preparation procedure (mass of the cosmetic product, solvent volume, volume of the supernatant, and reconstitution solvent). The remaining validation parameters, including intra- and inter-day accuracy and precision, injection repeatability, and stability were within the acceptance criteria (Table 3).

Table 3. Validation data.

	R	R-palm	R-ac	β-car	E	E-ac	Q10
Range (mg/L)	0.25–100	0.38–150	0.25–100	0.25–100	8.00–800	8.00–800	2.50–250
R^2	0.9996	1.0000	0.9998	0.9998	0.9999	0.9999	0.9999
LOD (ng/g)	1.88	2.12	1.52	55.0	97.6	85.8	12.3
LOQ (ng/g)	6.20	7.00	5.02	182	322	283	40.7
Intra-day accuracy (%)	101.5 ± 1.1	101.4 ± 0.7	101.4 ± 0.9	100.2 ± 1.7	100.6 ± 2.5	101.2 ± 2.0	100.3 ± 3.3
Inter-day accuracy (%)	99.8 ± 3.7	100.5 ± 4.3	102.2 ± 2.8	98.2 ± 3.2	96.9 ± 3.0	97.2 ± 2.4	107.0 ± 0.1
Intra-day precision (%)	0.88 ± 0.64	0.92 ± 0.70	0.90 ± 0.65	1.67 ± 0.72	1.48 ± 0.68	1.46 ± 0.75	1.31 ± 0.78
Inter-day precision (%)	1.46 ± 1.02	1.66 ± 0.56	3.40 ± 1.54	1.97 ± 1.08	1.39 ± 0.59	1.62 ± 0.82	1.24 ± 0.85
Injection repeatability (%)	0.20	0.08	0.12	0.11	0.09	0.15	0.35
Stability (%)	100.8 ± 0.4	100.5 ± 0.7	100.9 ± 0.4	101.0 ± 0.5	96.3 ± 2.5	101.8 ± 0.3	103.0 ± 0.8

The results for accuracy, precision, and stability are presented as an average of the three QC in triplicate ± standard deviation.

The sample preparation procedure was found repeatable, as the RSD between the triplicates of the same cosmetic product was <5% for all tested cosmetic products (Figures 3, 5 and 6). The average recoveries for all evaluated analytes, except for retinyl acetate, which was not found in any cosmetic product, were also within the acceptance criterion 100 ± 10%.

3.2. Overview of the Tested Cosmetics

Within this study, we evaluated a total of 73 anti-ageing facial cosmetic products, with vitamin A, E, and/or coenzyme Q10. Cosmetics in various formulations (day and night creams, serums, eye creams, anti-ageing concentrates, and a face tonic) and price ranges (Table 2) were included in this study. Among the three evaluated active compounds, vitamin E was the most common ingredient, labelled in ≈73% of the tested products, followed by vitamin A (≈51%) and coenzyme Q10 (≈42%). Approximately half of the tested products included only one of the tested active compounds in the ingredients list, ≈35% included two (vitamin A and E or vitamin E and coenzyme Q10), and ≈15% included all three evaluated active compounds (Figure 1). In total, 10 cosmetic products with quantitatively declared content (vitamin A in all cases) were included in the study (Figure 1). Cosmetic products with quantitatively declared vitamin E and coenzyme Q10 contents were not found on the Slovenian market. The analyzed cosmetic products labelled the presence of different vitamin A forms (retinol, retinyl palmitate, β carotene, hydroxypinacolone retinoate, retinyl propionate) and vitamin E forms (tocopherol or tocopheryl acetate) (Figure 1). Coenzyme Q10 was labelled in its oxidized form, ubiquinone, in all tested cosmetics. Different forms of the same active compound were also labelled in some tested cosmetic products. Both vitamin E forms were labelled in 19 cosmetic products, and two vitamin A forms (mostly retinol and retinyl palmitate) in 7 products.

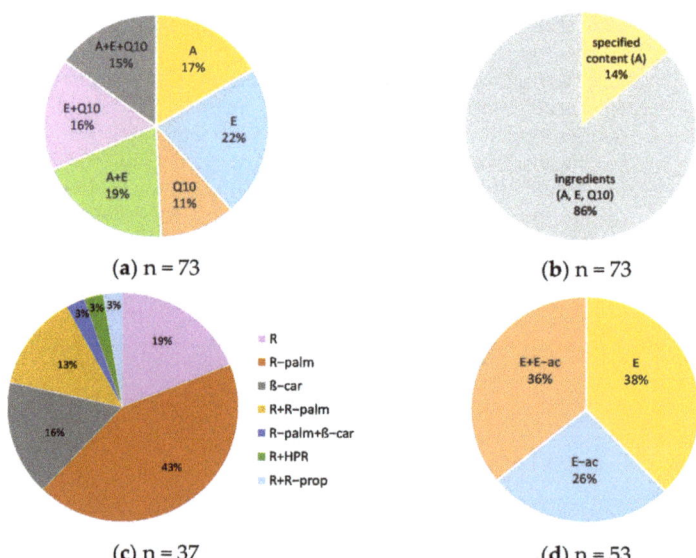

Figure 1. Distribution of the tested cosmetic products according to: (**a**) the labeled vitamins A and E and coenzyme Q10; (**b**) the quantitative specification of their content; (**c**) labeled vitamin A forms; (**d**) labeled vitamin E forms; n—number of the cosmetic products within each category.

3.3. Accuracy of the Labeling of Vitamins A and E and Coenzyme Q10

Within quality control of the tested cosmetics, we evaluated the accuracy of the labelling of vitamins A and E and coenzyme Q10. More specifically, we evaluated whether the labelled coenzyme Q10 and specific vitamin A and/or E forms are present in the cosmetics and whether the detected vitamin A and E forms are properly labelled. The results for all evaluated active compounds are summarized in Figure 2. Hydroxypinacolone retinoate was labelled and detected in one cosmetic product, as well as retinyl propionate. The accuracy of the labelling was also evaluated more comprehensively, considering the detected labelling errors in each of the tested cosmetic products. In total, at least one labelling error was observed in 31 cosmetic products (Figure 2). The observed labelling errors were uniformly distributed in cosmetic products purchased in different repositories and were observed in ≈40% of the cosmetics from pharmacies; ≈43% of the cosmetics from grocery stores and the Internet; and ≈44% of the cosmetics from drug stores.

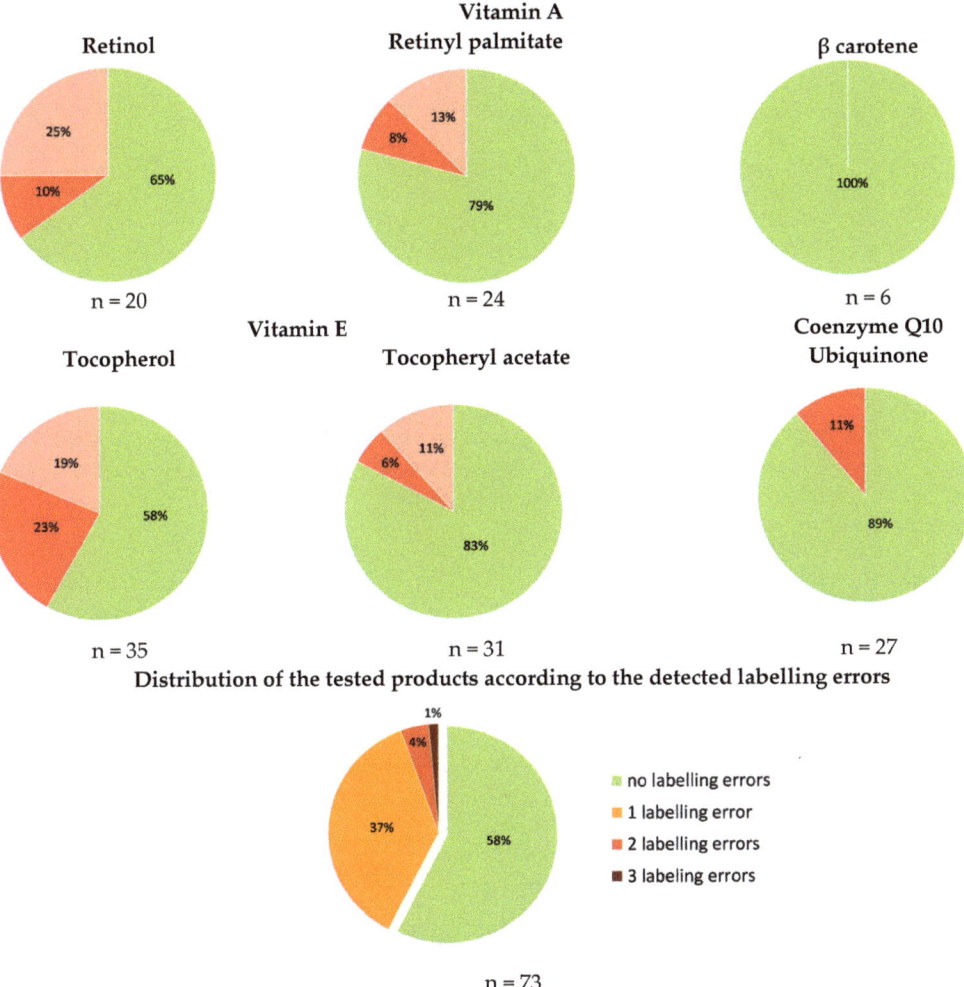

Figure 2. Accuracy of the labelling of vitamins A and E and coenzyme Q10 in the tested cosmetics (green—labelled and contained; pink—contained and not labelled; red—labelled and not contained) and the distribution of the tested products according to the detected labelling errors; n—number of tested cosmetic products within each category; labelling errors include the absence of the labelled (form of) compound and presence of a compound (in a form) which is not labelled.

3.4. Quantitative Evaluation of Vitamins A and E and Coenzyme Q10 in the Tested Cosmetics

The quality control of the tested cosmetic products also included quantification of the active compounds. The individual vitamin A forms (retinol, retinyl palmitate, β carotene, hydroxypinacolone retinoate, and retinyl propionate) were quantitatively determined in 35 different cosmetics which claimed their presence (Figure 3). The determined retinol concentrations ranged between 5.5 µg/g and 19 mg/g, with an average of 3.2 mg/g and a median of 390 µg/g. Retinyl palmitate was determined in concentrations between 4.0 µg/g and 9.2 mg/g, with an average of 1.0 mg/g and a median of 230 µg/g. The determined contents of β carotene ranged from 950 ng/g to 8.0 µg/g, with an average of 3.1 µg/g and a median of 2.1 µg/g. Hydroxypinacolone retinoate and retinyl propionate were each detected only once, both in cosmetic products which declared their presence. In cosmetics containing vitamin A not (only) in its retinol form, total vitamin A content, expressed with retinol equivalents (RE), was also determined (Figure 4).

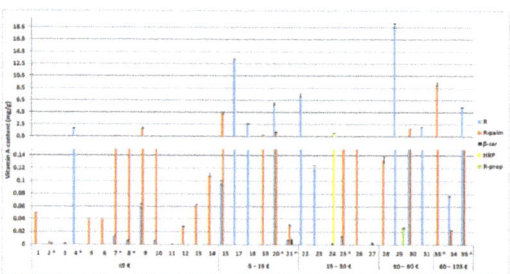

Figure 3. Determined content of retinol (R), retinyl palmitate (R-palm), β carotene (β-car), hydroxypinacolone retinoate (HRP), and retinyl propionate (R-prop) in the tested cosmetics (1–35), expressed in mg per g of the cosmetic product (average ± SEM, n = 3) in relation to the product's price (per 50 mL). In the tested cosmetics numbered 16 and 33, the labelled vitamin A was not detected. In the tested cosmetics marked with *, a labelling error was observed.

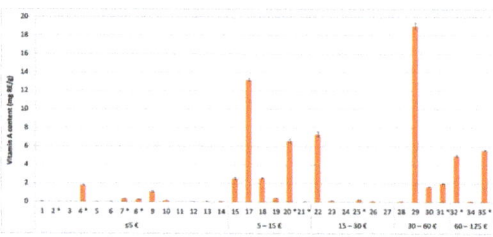

Figure 4. Determined total retinoid content, expressed in mg of retinol equivalents (RE) per g of the tested cosmetics (1–35) in relation to the product's price (per 50 mL). In the tested cosmetics numbered 16 and 33, the labelled vitamin A was not detected. In the tested cosmetics marked with *, a labelling error was observed.

The contents of tocopherol and tocopheryl acetate were determined in 49 cosmetics (Figure 5). Vitamin E as tocopherol or tocopheryl acetate was labelled as an ingredient in 48 evaluated cosmetics. One additional cosmetic product contained vitamin E which was not labelled. None of the tested cosmetics quantitatively declared the concentration of vitamin E. The determined tocopheryl acetate concentrations ranged between 35 µg/g and 16 mg/g, with an average of 5.5 mg/g and a median of 4.8 mg/g. Significantly lower tocopherol concentrations were generally determined in the tested cosmetics, ranging from 8.5 µg/g to 8.0 mg/g, with an average of 810 µg/g and a median of 120 µg/g.

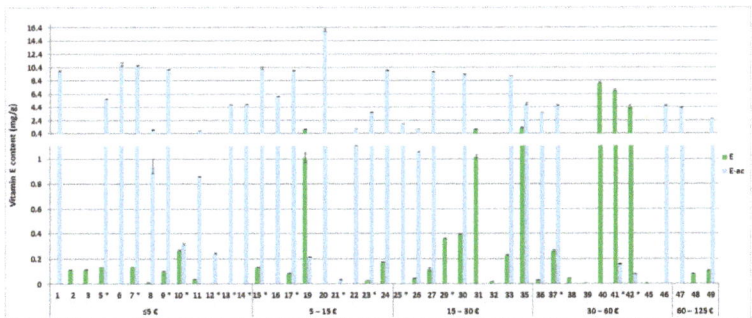

Figure 5. Determined content of tocopherol (E) and tocopheryl acetate (E-ac) in the tested cosmetics (1–49) in mg per g of the cosmetic product (average ± SEM, n = 3) in relation to the product's price (per 50 mL). In the tested cosmetics numbered 4, 18, 28, 34, 43, and 44, the labelled vitamin E was not detected. In the tested cosmetics marked with *, a labelling error was observed.

Coenzyme Q10, in its oxidized form, ubiquinone, was evaluated in 27 cosmetics which included it in the ingredients list. The content of ubiquinone was not stated in any of the tested cosmetics. The labelled ubiquinone was not detected in three cosmetics. The determined ubiquinone concentrations in the remaining 24 products ranged between 4.2 µg/g and 100 µg/g (Figure 6), with an average of 35 µg/g and a median of 25 µg/g.

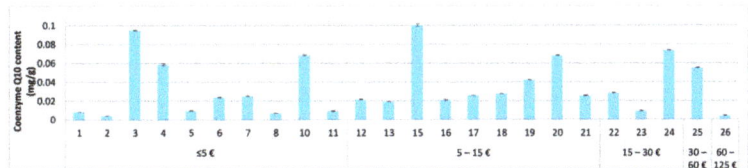

Figure 6. Determined content of coenzyme Q10 in the tested cosmetics (1–27) in mg per g of the cosmetic product (average ± SEM, n = 3) in relation to the product's price (per 50 mL). In the tested cosmetics numbered 9, 14, and 27, the labelled coenzyme Q10 was not detected.

3.5. Content-Related Quality Control of Vitamin A in the Tested Cosmetics

Ten of the tested cosmetic products quantitatively specified the content of a particular vitamin A form, which was most commonly retinol (eight products), as well as retinyl palmitate and hydroxypinacolone retinoate, each in one product. The obtained results on retinoid contents were compared to the label claims (Figure 7). Retinoid contents deviated significantly (by >20%) from the label claims in eight of the ten tested cosmetics. Retinoid contents ranged from 0% up to almost 400% of the label claims, with an average of 104% and a median of 95%.

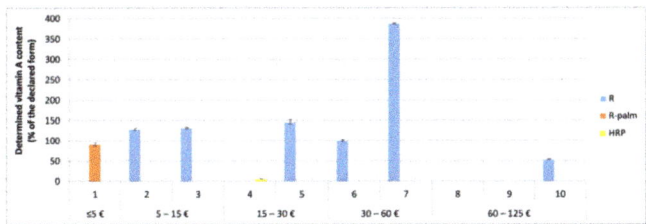

Figure 7. Determined vitamin A contents in relation to the label claim (%) in the 10 tested cosmetic products.

4. Discussion

Within this study, we evaluated the quality of 73 cosmetic products with vitamin A, E, and/or coenzyme Q10. Considering their beneficial effects against signs of photo ageing and intrinsic skin ageing [2,7], we focused on anti-ageing leave-on cosmetic products designed for facial use. Cosmetics in various formulations and with different labelled vitamin A and E forms (Table 2) were tested to provide representative samples for each of the three groups of active compounds and to obtain diversity in the prices, marketing, and accessibility of the products. The market survey confirmed that vitamin A, E, and coenzyme Q10 are widespread in cosmetic products. Among them, vitamin E was most commonly labelled in a variety of cosmetics products. Combinations of vitamin E and vitamin A and/or coenzyme Q10 were also commonly found on the market, while the combination of vitamin A and coenzyme Q10 without vitamin E was not found. This is also evident from the range of tested cosmetic products (Table 2 and Figure 1). The evaluated commercial cosmetics labelled the presence of different forms of vitamin A (mostly retinyl palmitate, retinol, and β carotene) and vitamin E (tocopherol and tocopheryl acetate), while coenzyme Q10 was only labelled in its oxidized form (ubiquinone) (Figure 1). Despite the lower activity than retinol [14,16], the more stable vitamin A form, retinyl palmitate [24], was the most frequently labelled vitamin A form. Newer vitamin A forms with higher activity and reduced incidence and intensity of irritation side effects are emerging on the market. Such examples are hydroxypinacolone retinoate [38] and retinyl propionate [39], each labelled in one cosmetic product. Vitamin E was more frequently labelled in the active form tocopherol, despite its lower stability [4]. However, tocopheryl acetate, individually or in combination with tocopherol, was also commonly found (Figure 1). An important selection criterion was also the specification of the active compounds' contents, which is less common in the cosmetic industry and was only found for vitamin A in 14% of the tested products (Table 2 and Figure 1). The quantitative specification of the active compounds' contents is a developing practice in recent years, especially in functional cosmetics, which promote different effects on the skin.

An appropriate, selective, and accurate methodology is a prerequisite for the quality control of cosmetics. The analysis of different vitamin A forms within this study was performed by a previously published HPLC–UV method for their quality control [34], which was selected for the comprehensive analysis of more retinoids. The simultaneous analysis of coenzyme Q10 and vitamin E, as tocopherol or tocopheryl acetate, was performed by a novel method for their quality control, comprising a simple sample preparation procedure and an HPLC–UV method for their quantification. Both methods utilized within this study were properly validated following the ICH guidelines [37] and proven suitable for the quality control of vitamins A and E and coenzyme Q10 in cosmetic products.

The quality control of the selected cosmetic products initially comprised the accuracy of the labelling of the evaluated active compound groups (different forms of vitamin A and E as well as coenzyme Q10). In general, the majority of the 35 evaluated vitamin A cosmetics contained some vitamin A form, except for two products. More inconsistencies were detected regarding the labelling of individual vitamin A forms. Retinol was not properly labelled in 35% and retinyl palmitate in 21% of the tested products (Figure 2). The labelling inconsistencies were mostly on account of their presence, which was not stated on the packaging, although their absence and replacement of the labelled retinyl palmitate with retinol were also noticed. No labelling inconsistencies were observed in the case of the less frequently found forms—β carotene, hydroxypinacolone retinoate, and retinyl propionate. These results are supported by our previous preliminary study on a smaller sample of retinoid cosmetics [34]. More labelling inconsistencies were observed in the case of vitamin E, which was not detected in ≈12% of the evaluated vitamin E cosmetics, and was present in one additional cosmetic product, which did not state it. Incorrect labelling of tocopheryl acetate was observed in 17% of the tested cosmetics, mostly due to its unlabeled presence (Figure 2). Contrary findings were obtained in the case of tocopherol, for which labelling inconsistencies were more frequently determined—in 42% of the tested cosmetics

(Figure 2). The observed coenzyme Q10 labelling inconsistencies in 11% of the tested cosmetics were a consequence of its absence. Although proper labelling of the ingredients is essential for quality assurance, there is a lack of studies researching the labelling accuracy of commercial cosmetics, including cosmetics with vitamin E and coenzyme Q10.

Further on, we evaluated the contents of different vitamin A and E forms, as well as coenzyme Q10 in the examined cosmetics. The determined vitamin A contents in the 35 tested cosmetic products varied greatly by a >4300-fold difference between the lowest and highest determined content for retinol, >2300-fold for retinyl palmitate, and >840-fold for β carotene (Figure 3). According to the literature, a significant facial anti-ageing effect may be achieved with topical formulations containing 0.075% [40] retinol or more (0.1% [41], 0.15% [42], 0.4% [43], 0.5% [44], and 1% [43]), whereas lower retinol concentration (0.04%) showed less prominent improvements of fine wrinkles, and no improvements of deep wrinkles [40]. Retinol contents near or above 0.075% were determined in half of the tested retinol cosmetics.

A more feasible approximation and prediction of the retinoid effects may be achieved by the determination of total vitamin A content expressed with retinol equivalents (Figure 4). Twelve (34%) of the tested cosmetics contained vitamin A in concentrations which are likely to achieve a significant anti-ageing effect (>0.075% RE). The efficacy of the tested cosmetics with >10-fold lower vitamin A contents (23% of the cosmetics) and >100-fold lower contents (23% of the cosmetics) is questionable. The remaining 20% of the tested vitamin A cosmetics contained >1000-fold less vitamin A than what is considered effective and are unlikely to achieve the desired anti-ageing effect. Most of these products belong to the lower price range ($\leq$ EUR 5/50 mL). However, considering their occurrence in each price range, except between EUR 30 and 60/50 mL, and the absence of a correlation between vitamin A content and the cosmetic products' price (Figure 4), we conclude that the price is not a determining factor for higher vitamin A content nor efficacy of the cosmetics. Another important aspect of vitamin A cosmetics is their safety, associated with their local adverse effects (potential retinoid-associated irritation and photo toxicity [14,45]) and systemic adverse effects (potential headaches, abdominal pain, nausea, liver or kidney damage, hypercalcemia, bone abnormalities, and teratogenicity [14,46]). Therefore, the Scientific Committee on Consumer Safety (SCCS), at the European Commission, recommends the use of vitamin A at maximum use concentrations of 0.3% RE in hand and face creams and other leave-on products [14]. The significantly higher total vitamin A content than the maximum recommended concentration by the SCCS in six of the 35 tested cosmetic products (Figure 4) is a cause of concern and reveals the need for (stricter) content-related quality control.

The determined vitamin E (tocopherol and tocopheryl acetate) contents showed lower variability than vitamin A contents by up to 500-fold for tocopheryl acetate and up to 1000-fold for tocopherol. The determined contents of tocopheryl acetate mostly gravitated towards 1% and were generally higher than the determined tocopherol contents, which were <0.1% in 90% of the tested products (Figure 5). The determined tocopherol and tocopheryl acetate contents were mostly within the expected range for facial leave-on cosmetics (0.03–2% for tocopherol and 0.003–6% for tocopheryl acetate), based on industry data on cosmetic products formulations [13] and research data (0.107–0.670% tocopheryl acetate in four commercial cosmetics on the Kuwait market [47]). Despite the differing information on tocopheryl acetate conversion to tocopherol (from 0% to 50%) found in the literature [10,48,49], the determined tocopherol and tocopheryl acetate contents were generally lower than the minimal effective tocopherol concentration of 1.0%, as recommended by Nada et al. [48]. A trend of increasing vitamin E content among the higher-priced products was not observed (Figure 5). Instead, vitamin E contents were more uniformly distributed between the different price ranges.

The tested commercial cosmetics showed the lowest variability, with only a 23-fold difference between the lowest and highest determined coenzyme Q10 content (Figure 6). The determined coenzyme Q10 contents are consistent with the survey data from the Voluntary

Cosmetic Registration Program (VCRP) for 2020 on a greater range of coenzyme Q10 leave-on cosmetics (387 products), with concentrations ranging between 0.00075% and 0.05% [22]. Coenzyme Q10 in concentrations ≥0.01% has shown beneficial anti-ageing effects on the skin [50], and were determined in <20% of the tested cosmetics. Additional research on the lowest effective coenzyme Q10 concentration in leave-on cosmetics is required to evaluate the efficacy of the remaining 80% of the tested coenzyme Q10 cosmetics. In three examined products, the labelled coenzyme Q10 was not detected. As previously observed for vitamins A and E, the expected correlation between the product prices and coenzyme Q10 content was not observed (Figure 6).

An important aspect within quality control is content-related quality control in relation to label claims, which is a generally accepted principle in the pharmaceutical industry, but has not yet been adopted in the cosmetics industry. One of the reasons, besides the looser regulation, is the fact that the active ingredients' contents in cosmetic products are rarely specified. This is also evident from Figure 1 as, despite being among the selection criteria, only 10 of the 73 tested cosmetics specified the content. The obtained results on the content in relation to the label claims (Figure 7) revealed significant deviations in both directions—from an absence or significantly lower content than declared up to 4-fold higher contents. Possible explanations for such deviations of the labelled vitamin A contents and the commonly determined active compounds contents below 0.01% include inappropriate formulation or their inappropriate stabilization and degradation during the manufacturing or storage [24]. Regardless, such results are concerning and support our recommendation for their stricter control and regulation, especially as the most significant deviations were observed in the higher-priced cosmetics (Figure 7).

5. Conclusions

Our work focused on the quality control of a significant number of cosmetic products with vitamins A and E and coenzyme Q10, which are common ingredients in anti-ageing cosmetics. On the example of these three groups of active compounds, we demonstrated an approach for the quality control of cosmetics, including evaluation of the labelling accuracy of different forms of active compounds, their content determination, and the content-related quality control in relation to the label claims. Based on the revealed labelling inconsistencies for all three groups of active compounds in 42% of the tested cosmetics, vitamin A contents above the maximum recommended concentration by the SCCS, and significant deviations in the contained and labelled vitamin A amounts, we recommend their stricter regulation and quality control. The development of suitable assay methods and progress in the field of functional cosmetics, which specify the content of active compounds, are essential steps towards their proper quality control following the principles of the pharmaceutic industry and the provision of quality, safe, and efficient cosmetics.

Supplementary Materials: The following are available online at https://www.mdpi.com/article/10.3390/cosmetics8030061/s1, Figure S1: Representative chromatogram of a standard mixture of retinol (retention time 2.5 min), retinyl acetate (retention time 2.9 min), and retinyl palmitate (retention time 6.9 min) at detection wavelength 325 nm, Figure S2: Representative chromatogram of β carotene standard solution (retention time 6.8 min) at detection wavelength 450 nm, Figure S3: Representative chromatograms of a standard mixture of tocopherol (retention time 6.0 min), tocopheryl acetate (retention time 6.4) and coenzyme Q10 (ubiquinone) (retention time 10.8) at detection wavelength 280 nm, Figure S4. Representative chromatograms of a cosmetic product with retinol (retention time 2.4) at detection wavelength 325 nm, Figure S5. Representative chromatograms of a cosmetic product with retinyl palmitate (retention time 6.9) at detection wavelength 325 nm, Figure S6. Representative chromatograms of a cosmetic product with β carotene (retention time 6.8) at detection wavelength 450 nm, Figure S7. Representative chromatograms of a cosmetic product with tocopherol (retention time 6.0 min), tocopheryl acetate (retention time 6.3) and coenzyme Q10 (ubiquinone) (retention time 10.7) at detection wavelength 280 nm.

Author Contributions: Conceptualization, R.R.; methodology, R.R.; validation, R.R. and Ž.T.R.; formal analysis, Ž.T.R.; resources, R.R.; data curation, Ž.T.R.; writing—original draft preparation,

Ž.T.R.; writing—review and editing, Ž.T.R. and R.R.; visualization, Ž.T.R.; supervision, R.R.; project administration, R.R.; funding acquisition, R.R. All authors have read and agreed to the published version of the manuscript.

Funding: This research was funded by Slovenian Research Agency (ARRS), grant number P1-0189.

Institutional Review Board Statement: Not applicable.

Informed Consent Statement: Not applicable.

Data Availability Statement: Data is contained within the article or supplementary material.

Conflicts of Interest: The authors declare no conflict of interest.

References

1. Lupo, M.P. Antioxidants and Vitamins in Cosmetics. *Clin. Dermatol.* **2001**, *19*, 467–473. [CrossRef]
2. Bissett, D.L. Common Cosmeceuticals. *Clin. Dermatol.* **2009**, *27*, 435–445. [CrossRef] [PubMed]
3. Thiele, J.J.; Hsieh, S.N.; Ekanayake-Mudiyanselage, S. Vitamin E: Critical Review of Its Current Use in Cosmetic and Clinical Dermatology. *Dermatol. Surg.* **2005**, *31*, 805–813. [CrossRef] [PubMed]
4. Thiele, J.J.; Ekanayake-Mudiyanselage, S. Vitamin E in Human Skin: Organ-Specific Physiology and Considerations for Its Use in Dermatology. *Mol. Asp. Med.* **2007**, *28*, 646–667. [CrossRef]
5. Afonso, S.; Horita, K.; Sousa e Silva, J.P.; Almeida, I.F.; Amaral, M.H.; Lobão, P.A.; Costa, P.C.; Miranda, M.S.; Esteves da Silva, J.C.G.; Sousa Lobo, J.M. Photodegradation of Avobenzone: Stabilization Effect of Antioxidants. *J. Photochem. Photobiol. B* **2014**, *140*, 36–40. [CrossRef]
6. Farris, P.; Yatskayer, M.; Chen, N.; Krol, Y.; Oresajo, C. Evaluation of Efficacy and Tolerance of a Nighttime Topical Antioxidant Containing Resveratrol, Baicalin, and Vitamin e for Treatment of Mild to Moderately Photodamaged Skin. *J. Drugs Dermatol.* **2014**, *13*, 1467–1472.
7. Masaki, H. Role of Antioxidants in the Skin: Anti-Aging Effects. *J. Dermatol. Sci.* **2010**, *58*, 85–90. [CrossRef]
8. Burke, K.E. Interaction of Vitamins C and E as Better Cosmeceuticals. *Dermatol. Ther.* **2007**, *20*, 314–321. [CrossRef]
9. Alberts, D.S.; Goldman, R.; Xu, M.J.; Dorr, R.T.; Quinn, J.; Welch, K.; Guillen-Rodriguez, J.; Aickin, M.; Peng, Y.M.; Loescher, L.; et al. Disposition and Metabolism of Topically Administered Alpha-Tocopherol Acetate: A Common Ingredient of Commercially Available Sunscreens and Cosmetics. *Nutr. Cancer* **1996**, *26*, 193–201. [CrossRef]
10. Baschong, W.; Artmann, C.; Hueglin, D.; Roeding, J. Direct Evidence for Bioconversion of Vitamin E Acetate into Vitamin E: An Ex Vivo Study in Viable Human Skin. *J. Cosmet. Sci.* **2001**, *52*, 155–161.
11. Rangarajan, M.; Zatz, J.L. Kinetics of Permeation and Metabolism of Alpha-Tocopherol and Alpha-Tocopheryl Acetate in Micro-Yucatan Pig Sin. *J. Cosmet. Sci.* **2001**, *52*, 35–50.
12. Nabi, Z.; Tavakkol, A.; Dobke, M.; Polefka, T.G. Bioconversion of Vitamin E Acetate in Human Skin. *Curr. Probl. Dermatol.* **2001**, *29*, 175–186.
13. Zondlo Fiume, M. Final Report on the Safety Assessment of Tocopherol, Tocopheryl Acetate, Tocopheryl Linoleate, Tocopheryl Linoleate/Oleate, Tocopheryl Nicotinate, Tocopheryl Succinate, Dioleyl Tocopheryl Methylsilanol, Potassium Ascorbyl Tocopheryl Phosphate, and Tocophersolan. *Int. J. Toxicol.* **2002**, *21*, 51–116.
14. Rousselle, C. Opinion of the Scientific Committee on Consumer Safety (SCCS)—Final Version of the Opinion on Vitamin A (Retinol, Retinyl Acetate and Retinyl Palmitate) in Cosmetic Products. *Regul. Toxicol. Pharmacol.* **2017**, *84*, 102–104. [CrossRef]
15. European Parliament and of the Council. Regulatution (EC) No 1223/2009 of the European Parliament and of the Council of 30 November 2009 on Cosmetic Products. *Off. J. Eur. Union* **2009**, *52*, L 342/83–L 342/127.
16. Zasada, M.; Budzisz, E. Retinoids: Active Molecules Influencing Skin Structure Formation in Cosmetic and Dermatological Treatments. *Postepy Dermatol. Alergol.* **2019**, *36*, 392–397. [CrossRef]
17. Temova Rakuša, Ž.; Srečnik, E.; Roškar, R. Novel HPLC-UV Method for Simultaneous Determination of Fat-Soluble Vitamins and Coenzyme Q10 in Medicines and Supplements. *Acta Chim. Slov.* **2017**, *64*, 523–529. [CrossRef]
18. Vinson, J.; Anamandla, S. Comparative Topical Absorption and Antioxidant Effectiveness of Two Forms of Coenzyme Q10 after a Single Dose and after Long-Term Supplementation in the Skin of Young and Middle-Aged Subjects. *Int J. Cosmet. Sci.* **2006**, *28*, 148. [CrossRef]
19. Knott, A.; Achterberg, V.; Smuda, C.; Mielke, H.; Sperling, G.; Dunckelmann, K.; Vogelsang, A.; Krüger, A.; Schwengler, H.; Behtash, M.; et al. Topical Treatment with Coenzyme Q10-Containing Formulas Improves Skin's Q10 Level and Provides Antioxidative Effects. *Biofactors* **2015**, *41*, 383–390. [CrossRef]
20. Hoppe, U.; Bergemann, J.; Diembeck, W.; Ennen, J.; Gohla, S.; Harris, I.; Jacob, J.; Kielholz, J.; Mei, W.; Pollet, D.; et al. Coenzyme Q10, a Cutaneous Antioxidant and Energizer. *Biofactors* **1999**, *9*, 371–378. [CrossRef]
21. Kaci, M.; Belhaffef, A.; Meziane, S.; Dostert, G.; Menu, P.; Velot, É.; Desobry, S.; Arab-Tehrany, E. Nanoemulsions and Topical Creams for the Safe and Effective Delivery of Lipophilic Antioxidant Coenzyme Q10. *Colloids Surf. B Biointerfaces* **2018**, *167*, 165–175. [CrossRef] [PubMed]

22. Expert Panel for Cosmetic Ingredient Safety Members. Safety Assessment of Ubiquinone Ingredients as Used in Cosmetics. Cosmetic Ingredient Review. Available online: https://www.cir-safety.org/supplementaldoc/safety-assessment-ubiquinone-ingredients-used-cosmetics (accessed on 10 April 2021).
23. Temova Rakuša, Ž.; Kristl, A.; Roškar, R. Stability of Reduced and Oxidized Coenzyme Q10 in Finished Products. *Antioxidants* **2021**, *10*, 360. [CrossRef] [PubMed]
24. Rakuša, Ž.T.; Škufca, P.; Kristl, A.; Roškar, R. Retinoid Stability and Degradation Kinetics in Commercial Cosmetic Products. *J. Cosmet. Dermatol.* **2021**, *20*, 2350–2358. [CrossRef] [PubMed]
25. Benevenuto, C.G.; Guerra, L.O.; Gaspar, L.R. Combination of Retinyl Palmitate and UV-Filters: Phototoxic Risk Assessment Based on Photostability and in Vitro and in Vivo Phototoxicity Assays. *Eur. J. Pharm. Sci.* **2015**, *68*, 127–136. [CrossRef]
26. Carlotti, M.E.; Rossatto, V.; Gallarate, M.; Trotta, M.; Debernardi, F. Vitamin A Palmitate Photostability and Stability over Time. *J. Cosmet. Sci.* **2004**, *55*, 233–252. [CrossRef]
27. Guaratini, T.; Gianeti, M.D.; Campos, P.M.B.G.M. Stability of Cosmetic Formulations Containing Esters of Vitamins E and A: Chemical and Physical Aspects. *Int. J. Pharm.* **2006**, *327*, 12–16. [CrossRef]
28. Gianeti, M.D.; Gaspar, L.R.; Bueno de Camargo Júnior, F.; Campos, P.M.B.G.M. Benefits of Combinations of Vitamin A, C and E Derivatives in the Stability of Cosmetic Formulations. *Molecules* **2012**, *17*, 2219–2230. [CrossRef]
29. Moyano, M.A.; Segall, A. Vitamin a Palmitate and α-Lipoic Acid Stability in O/W Emulsions for Cosmetic Application. *J. Cosmet. Sci.* **2011**, *62*, 405–415.
30. Akhavan, A.; Levitt, J. Assessing Retinol Stability in a Hydroquinone 4%/Retinol 0.3% Cream in the Presence of Antioxidants and Sunscreen under Simulated-Use Conditions: A Pilot Study. *Clin. Ther.* **2008**, *30*, 543–547. [CrossRef]
31. Ceugniet, C.; Lepetit, L.; Viguerie, N.L.D.; Jammes, H.; Peyrot, N.; Rivière, M. Single-Run Analysis of Retinal Isomers, Retinol and Photooxidation Products by High-Performance Liquid Chromatography. *J. Chromatogr. A* **1998**, *810*, 237–240. [CrossRef]
32. Wang, L.-H.; Huang, S.-H. Determination of Vitamins A, D, E, and K in Human and Bovine Serum, and B-Carotene and Vitamin A Palmitate in Cosmetic and Pharmaceutical Products, by Isocratic HPLC. *Chromatographia* **2002**, *55*, 289–296. [CrossRef]
33. Gatti, R.; Gioia, M.G.; Cavrini, V. Analysis and Stability Study of Retinoids in Pharmaceuticals by LC with Fluorescence Detection. *J. Pharm. Biomed. Anal.* **2000**, *23*, 147–159. [CrossRef]
34. Temova Rakuša, Ž.; Škufca, P.; Kristl, A.; Roškar, R. Quality Control of Retinoids in Commercial Cosmetic Products. *J. Cosmet. Dermatol.* **2021**, *20*, 1166–1175. [CrossRef]
35. Hubinger, J.C. Determination of Retinol, Retinyl Palmitate, and Retinoic Acid in Consumer Cosmetic Products. *J. Cosmet. Sci.* **2009**, *60*, 485–500. [CrossRef]
36. Grace, A.C.; Prabha, T.; Jagadeeswaran, M.; Srinivasan, K.; Sivakumar, T. Analytical Method Development for Simultaneous Determination of Ubidecarenone and Vitamin E Acetate in Capsule Dosage Form by HPLC. *Int. J. Pharm. Pharm. Sci.* **2019**, 79–84. [CrossRef]
37. ICH Q2(R1). Validation of Analytical Procedures: Text and Methodology. ICH Harmonised Tripartite Guideline Harmonised Tripartite Guideline. 2005. Available online: http://www.ich.org/products/guidelines/quality/quality-single/article/validation-of-analytical-procedures-text-and-methodology.html (accessed on 11 June 2018).
38. Ruth, N.; Mammone, T. Anti-Aging Effects of Retinoid Hydroxypinacolone Retinoate on Skin Models. *J. Invest. Dermatol.* **2018**, *138*, S223. [CrossRef]
39. Bjerke, D.L.; Li, R.; Price, J.M.; Dobson, R.L.M.; Rodrigues, M.; Tey, C.; Vires, L.; Adams, R.L.; Sherrill, J.D.; Styczynski, P.B.; et al. The Vitamin A Ester Retinyl Propionate Has a Unique Metabolic Profile and Higher Retinoid-Related Bioactivity over Retinol and Retinyl Palmitate in Human Skin Models. *Exp. Dermatol.* **2021**, *30*, 226–236. [CrossRef]
40. Kikuchi, K.; Suetake, T.; Kumasaka, N.; Tagami, H. Improvement of Photoaged Facial Skin in Middle-Aged Japanese Females by Topical Retinol (Vitamin A Alcohol): A Vehicle-Controlled, Double-Blind Study. *J. Dermatol. Treat.* **2009**, *20*, 276–281. [CrossRef]
41. Kong, R.; Cui, Y.; Fisher, G.J.; Wang, X.; Chen, Y.; Schneider, L.M.; Majmudar, G. A Comparative Study of the Effects of Retinol and Retinoic Acid on Histological, Molecular, and Clinical Properties of Human Skin. *J. Cosm. Dermatol.* **2016**, *15*, 49–57. [CrossRef]
42. Zasada, M.; Budzisz, E. Randomized Parallel Control Trial Checking the Efficacy and Impact of Two Concentrations of Retinol in the Original Formula on the Aging Skin Condition: Pilot Study. *J. Cosmet. Dermatol.* **2020**, *19*, 437–443. [CrossRef]
43. Kafi, R.; Kwak, H.S.R.; Schumacher, W.E.; Cho, S.; Hanft, V.N.; Hamilton, T.A.; King, A.L.; Neal, J.D.; Varani, J.; Fisher, G.J.; et al. Improvement of Naturally Aged Skin with Vitamin A (Retinol). *Arch. Dermatol.* **2007**, *143*, 606–612. [CrossRef]
44. Gold, M.H.; Kircik, L.H.; Bucay, V.W.; Kiripolsky, M.G.; Biron, J.A. Treatment of Facial Photodamage Using a Novel Retinol Formulation. *J. Drugs. Dermatol.* **2013**, *12*, 533–540.
45. Tolleson, W.; Cherng, S.-H.; Xia, Q.; Boudreau, M.; Yin, J.; Wamer, W.; Howard, P.; Yu, H.; Fu, P. Photodecomposition and Phototoxicity of Natural Retinoids. *Int. J. Environ. Res. Public Health* **2005**, *2*, 147–155. [CrossRef]
46. Rutkowski, M.; Grzegorczyk, K. Adverse Effects of Antioxidative Vitamins. *Int. J. Occup. Med. Environ. Health* **2012**, *25*, 105–121. [CrossRef]
47. Nada, A.; Krishnaiah, Y.S.R.; Zaghloul, A.-A.; Khattab, I. Analysis of Vitamin E in Commercial Cosmetic Preparations by HPLC. *J. Cosmet. Sci.* **2010**, *61*, 353–365.
48. Nada, A.; Krishnaiah, Y.S.R.; Zaghloul, A.-A.; Khattab, I. In Vitro and in Vivo Permeation of Vitamin E and Vitamin E Acetate from Cosmetic Formulations. *Med. Princ. Pract.* **2011**, *20*, 509–513. [CrossRef]

49. Trevithick, J.R.; Mitton, K.P. Topical Application and Uptake of Vitamin E Acetate by the Skin and Conversion to Free Vitamin E. *Biochem. Mol. Biol. Int.* **1993**, *31*, 869–878.
50. Prahl, S.; Kueper, T.; Biernoth, T.; Wöhrmann, Y.; Münster, A.; Fürstenau, M.; Schmidt, M.; Schulze, C.; Wittern, K.-P.; Wenck, H.; et al. Aging Skin Is Functionally Anaerobic: Importance of Coenzyme Q10 for Anti Aging Skin Care. *Biofactors* **2008**, *32*, 245–255. [CrossRef]

Article

Impact of Solar Ultraviolet Radiation in the Expression of Type I Collagen in the Dermis

Foteini Biskanaki [1,*], Efstathios Rallis [1], George Skouras [2], Anastasios Stofas [3], Eirini Thymara [3], Nikolaos Kavantzas [3], Andreas C. Lazaris [3] and Vasiliki Kefala [1]

[1] Department of Biomedical Sciences, School of Health Sciences and Welfare, University of West Attica, 12243 Athens, Greece; erallis@uniwa.gr (E.R.); valiakef@uniwa.gr (V.K.)
[2] Department of Plastic Surgery, Skouras Med Clinics, 10673 Athens, Greece; giorgos_skouras@yahoo.gr
[3] First Department of Pathology, School of Medicine, National and Kapodistrian University of Athens, 11527 Athens, Greece; astofas@otenet.gr (A.S.); ei_thymara@yahoo.gr (E.T.); nkavantz@med.uoa.gr (N.K.); alazaris@med.uoa.gr (A.C.L.)
* Correspondence: fbiskanaki@uniwa.gr; Tel.: +30-6988132130

Abstract: Ultraviolet radiation exposure is the dominant environmental determinant of all major forms of skin cancer, and the main cause of prematurely aged skin that is referred to as photoaging. Collagen type I (COL I) is expressed differently along with the dermis between healthy and pathological skin tissues. The aim of this study was to understand the impact of solar radiation in the dermis and assess the impact of solar radiation to COL I. The hematoxylin and eosin staining protocol was performed in tissue paraffin blocks and then they were stained immunohistochemically with the rabbit monoclonal anti-COL I antibody. A total of 270 slides were studied with an Olympus BX 41 microscope; we scored positively the expression of COL I in dermis and statistically analyzed with IBM SPSS Statistics. Based on our results, we observed that solar elastosis changes the structure of the skin's collagen. In healthy tissues, COL I had a uniform expression along with the dermis. In tissues with aging, COL I expression was weaker and lost homogeneity. In pathological tissues (non-melanoma skin cancers, NMSCs), precancerous lesions, and benign skin lesions), the expression of COL I was observed to be almost weaker than tissues with aging in all body parts and much weaker below the lesions. The most severe solar elastosis was observed in the extremities. The degree of severity of the solar elastosis in relation to age did not appear to be completely affected. Solar radiation divides the collagen more rapidly than normal biological aging and solar elastosis was observed into the skin tissues with photoaging, which replaces the collagen fibers of the skin. These results confirm previous studies, which have shown that skin COL I decreases during aging, more in photoaging and even more in skin cancers. We conclude that skin COL I expression is reduced as a result of ultraviolet radiation and leading to negative impacts on the skin.

Keywords: solar elastosis; collagen type I; solar radiation; non-melanoma skin cancers; photoaging

1. Introduction

It is now known that exposure to solar radiation can cause negative effects on the skin and human health. Sun damage is accumulative, so even a short exposure to the sun is added to the skin throughout a person's life. The skin is a vital organ that permits the body's communication with the environment. Radiation alters normal skin [1]. Ultraviolet radiation exposure is the dominant environmental determinant of all major forms of skin cancer, and the main cause of prematurely aged skin that is referred to as photoaging. Photoaging is also called actinic aging and can be caused by the breakdown of collagen, the formation of free radicals, and the interaction of DNA repair mechanisms and their inhibitory effect on immune mechanisms [2].

Solar elastosis is a degenerative condition of elastic tissue in the dermis due to prolonged sun exposure. There are a variety of clinical manifestations of solar elastosis;

most commonly appearing as yellow, thickened, and coarsely wrinkled skin [3]. Solar elastosis and the degeneration of collagen can be observed histologically using hematoxylin and eosin staining (H&E) [4]. These changes are due to an imbalance between the production and degradation of the main proteins produced by fibroblasts [2]. Among these proteins, the most important is type I collagen (COL I, fibrillar). Total skin collagen is made of 80 to 85% of COL I [5].

Skin aging (biological aging and photoaging) is caused by both endogenous and exogenous factors [6]. Endogenous aging is a process that leads to thin, dry skin with fine wrinkles and gradual skin atrophy. [7] Exogenous aging is caused by environmental factors such as air pollution, smoking, poor nutrition and sun exposure, resulting in rough wrinkles, loss of elasticity, relaxation, and a rough look [8]. Ultraviolet (UV) radiation causes oxidative stress in skin cells, resulting in damaged cells with oxidized lipids activating complement systems and causing inflammation, leading to infiltration and activation of macrophages. Activated macrophages release uterine metalloproteinases (MMPs) which break down the extracellular matrix [9]. Repeated ultraviolet radiation inactivates the complement system, causing damage to the epidermis–dermis junction, in which macrophages are deposited and are overloaded with oxidized lipids. Overloaded macrophages release pro-inflammatory cytokines and reactive oxygen species (ROS), which cause chronic inflammation and long-term damage to the dermis [10].

Skin cancers represent the most common type of cancer worldwide. Non-melanoma skin cancer (NMSC) refers to a group of cancers that slowly develop in the upper layers of the skin [11]. The term non-melanoma distinguishes these more common types of skin cancer (99% are basal cell carcinomas, BCCs) and squamous cell carcinomas (SCCs) from less common skin cancers such as melanoma [12].

This study is based on the different expressions of COL I in the dermis between healthy and pathological tissues (e.g., aging, solar elastosis, NMSC, etc.). The aim was to assess the impact of solar radiation on COL I.

2. Materials and Methods

2.1. Tissue Samples

Biopsies of severe sun damaged skin (n = 135) recovered from the First Department of Pathology of Medicine School of the National and Kapodistrian University of Athens in Greece. Tissue samples (n = 88, NMSC, and n = 47, healthy skins that were used as controls) were fixed in buffered formalin, embedded into paraffin blocks, and then stained with hematoxylin and eosin.

2.2. Antibodies

The rabbit monoclonal anti-COL I antibody [EPR7785] IHC-P 1/1500 was used. It was performed using heat-mediated antigen retrieval with Thermo Scientific Pierce Tris-EDTA (TE) buffer, pH 9, before commencing with IHC staining for protocollagen.

2.3. Immunohistochemistry Microscopy Analysis

The microscope slides were evaluated by using an Olympus BX 41 microscope in magnification ×40 and ×100. The immunohistochemical report was performed by estimating with visual evaluation the percentage of COL I expression on a scale of 1 to 5, positively (weak +, weak to moderate ++, moderate +++, moderate to severe ++++, and severe +++++) [5].

2.4. Statistical and Data Analysis

All the data collected were entered into an electronic database created by Excel software. Data analysis was performed using IBM SPSS Statistics for Windows, version 26.0. Frequencies were calculated for qualitative variables. Categorical variables were gender, age categories, body part, and type of lesion. They were studied using chi-square ($\times$2) and descriptive analysis, in relation to: (a) type of lesion, body part; (b) expression of COL I;

and (c) the degree of severity of solar elastosis. One sample *t*-test was applied to determine the different expression of COL I in sun-damaged skins. The Kolmogorov–Smirnov test was applied to check normality. This relationship was accessed by the Kruskal Wallis test, providing the mean and standard deviation. Values of $p < 0.05$ were indicative of statistical significance.

3. Results

3.1. Characteristics of Tissue Samples

Healthy tissue samples (n = 47) were from, the abdomen (n = 4), face (n = 22), and breast (n = 21). In terms of the pathological specimens (n = 88), 40 were from the face, 14 from the back, 12 from the abdomen, and 22 from extremities. A total of 44 of them had aging and 3 were youth skin. Of the 88 pathological tissues (42 male, 46 female), 86 had solar elastosis and 66 of them had more lesions, concurrently. A total of 23 of the 66 had been diagnosed as benign lesions (seborrheic keratosis and nevus), 3 as precancerous skin lesions (dysplastic nevus and actinic keratosis), and 38 as NMSC (basal cell carcinomas and squamous cell carcinomas). The specimens were divided into 3 age groups (1st group = 66–85 years old, 2nd = 46–65 years old, 3rd = 25–45 years old). The largest specimen in our study with NMSC (n = 74) concerned the age group of 66–85. The study focused on COL I's expression in three indexes. The first index was between the epidermis and solar elastosis (index A), the second was along the dermis (index B), and the last was below the cutaneous lesion (index C).

3.2. Healthy Tissue Samples

The results from the IHC microscopy analysis showed that the healthy skin samples had a uniform expression of COL I in the dermis. The expression of COL I in the healthy tissue samples with biological aging was uniform along the dermis and weaker than the expression of young skin. Moderate to intense (=4) expression was observed in the age group of 25–45, and moderate expression (=3) in the age group of 46–65, with a percentage of 65.96%. However, in chronological aging (in the age group of 66–85 years old), COL I's I expression was moderate (=3) and a small percentage (2.65%) showed a weak (=1) expression (ages over 75 years old).

COL I staining confirmed that the collagen fibers were thin and loose in the papillary dermis and thicker in the reticular dermis. The healthy samples with youthful skin and chronological ageing appeared with collagen fibers that were thin and loose in the papillary dermis and were thicker in the reticular dermis. The distance of collagen fibers was bigger from each other in samples with ageing, compared with youthful skin samples. In the aging tissues, the keratin layer showed hyperplasia, skin atrophy, and reduction of the number of skin components except for sebaceous glands that were overgrown. Weaker expression of COL I was generally observed in relation to the skin at a younger age.

3.3. Photoaging

The specimens with photoaging were assessed according to the severity of the solar elastosis per body part. Then, it was compared with the degree of COL I expression. The results of the average expression of COL I per age group and body part are delineated in Table 1.

In tissue samples with photoaging, the formation of a solar elastosis islet of elastin was observed beneath the skin, which replaced collagen. The average COL I expression between the epidermis and solar elastosis (index A) was weak to moderate and weak along with the dermis. Below the epidermis, it was observed that COL I was not expressed at all. The severest solar elastosis was observed in the extremities, then in the back, and less in the abdomen and face. Solar elastosis represented as a film-like distribution, except for four specimens with a very weak expression of COL I that was interrupted.

The degree of solar elastosis had a negative correlation with COL I index A, of the order of 42.3% (when the degree of elastosis increases by one unit, the effect of collagen decreases by 0.42 of the unit and vice versa; when the degree of elastosis decreases by one unit the effect of collagen increases by 0.42 of the unit).

Table 1. Average of the degree severity of solar elastosis per body part and average expression of COL I along the dermis (per body part, Index A, Index B).

Body Parts	Avg. Solar Elastosis	Avg. COL I (Index A)	Avg. COL I (Index B)
Hands	3.57	2.14	1.00
Legs	3.33	1.83	1.18
Abdomen	2.75	2.67	1.33
Thighs	1.50	2.00	1.00
Face	2.90	2.21	1.14
Back	3.50	2.29	1.08
Average	3.06	2.21	1.15

Moreover, a negative correlation was observed with COL I index B, of the order of 16% (which means that when, for example, the degree of elastosis increases by one unit, the effect of collagen decreases by 0.16 of the unit and vice versa; when the degree of elastosis is reduced by one unit the effect of collagen increases by 0.16 of the unit).

Of the statistical analysis comparison of COL I's expression in biological aging and photoaging, per age groups in the face to criterion B, the following was observed: In group 46–65 with biological aging, the expression of COL I was on average moderate; in specimens with photoaging, COL I's expression was weaker than biological aging. It was observed that in ages over 75 years old, solar elastosis was milder than at the age of 65 years old (Figure 1).

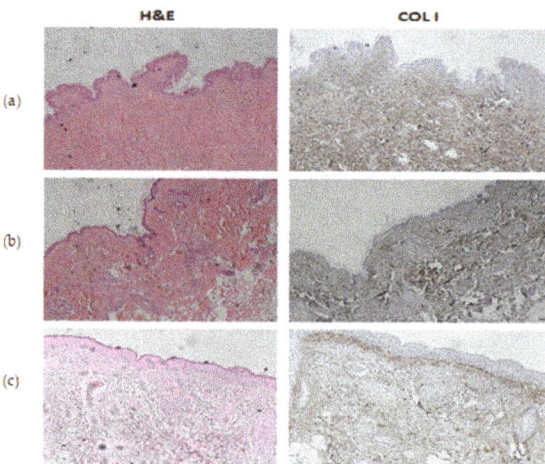

Figure 1. (a) Histochemical staining of hematoxylin–eosin and anti-COL I in youthful tissue. (b) H&E and anti-COL I staining: biological aging. (c) H&E and anti-COL I staining: photoaging.

3.4. Non-Melanoma Skin Cancers

The results in benign lesions were with an average of COL I expression almost weak to moderate (index A = 1.82, index C = 1.85) in all areas of the body. The average expression of COL I along with the dermis (index B) was weak. Our sample number regarding the precancerous lesions were limited, thus, the results are under consideration. However, it was observed that the average of COL I was moderately expressed in index A and weakly in indexes B and C. The average expression of COL I in the NMSCs was weakly expressed along with the dermis (index B) and weaker below the lesions (index C), while it was weak to moderate between the epidermis and solar elastosis (index A) (Figure 2).

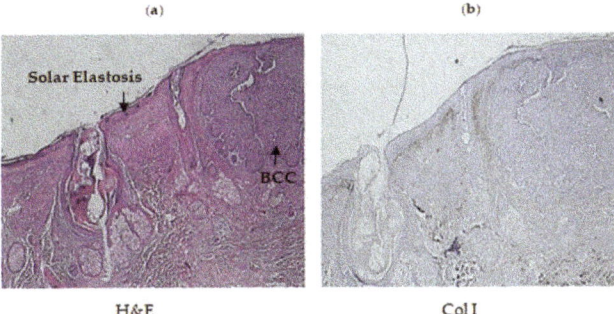

Figure 2. (**a**) Histochemical staining of hematoxylin–eosin in pathological tissue from the hand of a 77 year old woman. She had been diagnosed with solar elastosis and basal cell carcinoma (BCC). (**b**) Immunohistochemistry: anti-Col I antibody. Absence of COL I in the positions of solar elastosis. Moderate expression of COL I observed between the epidermis and solar elastosis (film-like distribution) which was weak in the rest of the dermis. Absence of expression observed below the lesion.

The average expression of COL I was weak in almost all body parts, and the abdomen had the maximum expression of COL I compared with skin tissues from other body areas (average 2.67 for index A, 1.33 for index B).

4. Discussion

To understand the impact of solar radiation in the dermis and assess the impact of solar radiation on COL I, we studied biopsies from healthy and pathological tissues and assessed the expression of COL I in these samples. In healthy tissues, COL I staining confirmed that the collagen fibers were thin and loose in the papillary dermis and thicker with homogeneity. However, with aging they became weaker and lost their homogeneity [7].

UVA radiation is absorbed in a percentage of 20% by the dermis and 80% by the epidermis. Thus, solar elastosis appears superficially and can change the structure of collagen and elastin fibers in the skin [8]. The effect of sunlight on the dermis causes an increase in elastin in quantity and MMPs are produced in large quantities [13]. Under normal conditions, these enzymes repair the "wound" from the sun-damaged crust, making and reconstituting collagen. This process is not always 100% successful and some MMPs breaks down collagen, producing decomposed collagen fibers, resulting in "solar scars" [7,14]. As well as direct UVA irradiation in the dermis, UVB-irradiated keratinocytes can affect collagen formation (degradation) in dermal fibroblasts through secretory factors such as inflammatory cytokines including interleukin-1 (IL-1) and tumor necrosis factor-α (TNF-α) in the skin. TNF-α stimulates the chemotaxis of inflammatory cells to the skin and downregulates procollagen mRNA, and thus a blockade may be beneficial to the production of type I collagen [13].

In our specimens, it was observed that the solar elastosis was more severe in the extremities, back, and less in the face. The severity of the solar elastosis in relation to age did not appear to be completely affected. It was observed that in people over the age of 75, solar elastosis was milder than in the age of 65, and we would expect the reduction to be more severe in older skin. This could be a random finding observed in our samples owing to various factors, for example, lifestyle, location, duration of sun exposure, etc. [5]. Nevertheless, the possible cause could be the relation with the ozone layer, as the stratospheric ozone is an effective UV absorber. As the ozone layer becomes thinner, the protective filter provided by the atmosphere gradually decreases. As a result, every year humans and the environment are exposed to higher levels of UV radiation with more severe adverse effects at younger ages than in the past [5]. In pathological tissues of our study (NMSCs, precancerous lesions) and in benign skin lesions, the expression of COL I was almost weaker than skin tissues with aging in all body parts and much weaker below the lesion.

Fligiel, S.E., et.al. found collagen changes in photodamaged skin and changes in collagen structure in aged and photodamaged skin. They suggested that collagen fragmentation in vivo could underlie the loss of collagen synthesis in photodamaged skin and, to a lesser extent perhaps, in aged skin [15]. Solar radiation divides the collagen more rapidly than normal biological aging [15]. Solar elastosis was observed in the skin samples with photoaging, which replaced the collagen fibers of the skin [5,6]. Our results confirmed previous reports, which showed that in photodamaged skin COL I decreases and solar elastosis changes the structure of the skin's collagen. In healthy tissues, COL I had a uniform expression in the dermis. In tissues with aging, COL I expression was weaker and lost homogeneity.

To our knowledge, this is the most multitudinous study in current literature to assess the impact of solar ultraviolet radiation in the expression of COL I in the dermis and compare its expression between healthy youth skin, aging, photoaging, benign skin lesions, and NMSCs. In conclusion, skin COL I expression is reduced as a result of ultraviolet radiation, which leads to negative impacts on the skin. COL I decreases during aging, more in photoaging, and even more in skin cancers.

Author Contributions: Conceptualization, F.B. and V.K.; methodology, A.C.L.; software, E.R.; validation, G.S., E.T. and N.K.; formal analysis, F.B.; investigation, F.B.; resources, A.S.; data curation, G.S.; writing—original draft preparation, F.B.; writing—review and editing, E.R. and A.S.; visualization, N.K.; supervision, A.C.L. and V.K.; and project administration; V.K. All authors have read and agreed to the published version of the manuscript.

Funding: This research received no external funding.

Institutional Review Board Statement: Not applicable.

Informed Consent Statement: Not applicable.

Data Availability Statement: Not applicable.

Acknowledgments: We thank A. Kapranou and V. Papadopoulos for their help in IHC analysis.

Conflicts of Interest: The authors declare no conflict of interest.

References

1. Nguyen, A.V.; Soulika, A.M. The Dynamics of the Skin's Immune System. *Int. J. Mol. Sci.* **2019**, *20*, 1811. [CrossRef] [PubMed]
2. Lai-Cheong, J.E.; McGrath, J.A. Structure and function of skin, hair and nails. *Medicine* **2017**, *45*, 347. [CrossRef]
3. Heng, J.K.; Aw, D.C.W.; Tan, K.B. Solar Elastosis in Its Papular Form: Uncommon, Mistakable. *Case Rep. Dermatol.* **2014**, *6*, 124–128. [CrossRef]
4. Knapp, M.; Carpenter, C.E.; Shea, K.; Stowman, A.; Pierson, J. Focal Dermal Elastosis. A Proposed Update to the Nomenclature. *Am. J. Dermatopathol.* **2020**, *42*, 774–775. [CrossRef] [PubMed]
5. Pain, S.; Berthélémy, N.; Naudin, C.; Degrave, V.; André-Frei, V. Understanding Solar Skin Elastosis-Cause and Treatment. *J. Cosmet. Sci.* **2018**, *69*, 175–185. [PubMed]
6. Weihermann, A.C.; Lorencini, M.; Brohem, C.A.; de Carvalho, C.M. The structure of elastin and its involvement in photosynthesis of the skin. *Int. J. Cosmet. Sci. Investig.* **2017**, *39*, 241–247. [CrossRef]
7. Zhang, S.; Duan, E. Fighting against Skin Ageing. The Way from Bench to Bedside. Available online: https://doi.org/10.2147/CCID.S191935 (accessed on 25 April 2018).
8. Varvaresou, A.; Tsirivas, E.; Tsaoula, E.; Protopapa, E. Oxidative stress, photoageing and topical antioxidant protection. *Rev. Clin. Pharmacol. Pharmacokinet.* **2014**, *18*, 261–266.
9. Shao, Y.; Qin, Z.; Alexander Wilks, J.; Balimunkwe, R.M.; Fisher, G.J.; Voorhees, J.J.; Quan, T. Physical properties of the photodamaged human skin dermis: Rougher Colsurface and stiffer/harder mechanical properties. *Exp. Dermatol.* **2019**, *28*, 914–921. [CrossRef] [PubMed]

10. Trudel, D.; Desmeules, P.; Turcotte, S.; Plante, M.; Grégoire, P.; Renaud, M.C.; Orain, M.; Bairati, I.; Têtu, B. Visual and automated assessment of matrix metalloproteinase-25 tissue expression for the evaluation of ovarian cancer prognosis. *Modern Pathol.* **2014**, *27*, 1394–1404. [CrossRef]
11. Saggini, A.; Cota, C.; Lora, V.; Kutzner, H.; Rütten, A.; Sangüeza, O.; Requena, L.; Cerroni, L. Uncommon Histopathological Variants of Malignant Melanoma. Part 2. *Am. J. Dermatopathol.* **2019**, *41*, 321–342. [CrossRef] [PubMed]
12. Apalla, Z.; Nashan, D.; Weller, R.B.; Castellsagué, X. Skin Cancer: Epidemiology, Disease Burden, Pathophysiology, Diagnosis, and Therapeutic Approaches. *Dermatol. Ther.* **2017**, *7*, 5–19. [CrossRef] [PubMed]
13. Sharma, M.R.; Mitrani, R.; Wertha, P. Effect of TNFα blockade on UVB-induced inflammatory cell migration and collagen loss in mice. *J. Photochem. Photobiol. Biol.* **2020**, *213*, 112072. [CrossRef] [PubMed]
14. Gregorio, I.; Braghetta, P.; Bonaldo, P.; Cescon, M. ColVI in healthy and diseased nervous system. *Cell Transplant.* **2018**, *27*, 729–738.
15. Fligiel, S.E.; Varani, J.; Datta, S.C.; Kang, S.; Fisher, G.J.; Voorhees, J.J. Col Degradation in Aged/Photodamaged Skin In Vivo and After Exposure to Matrix Metalloproteinase-1 In Vitro. *J. Investig. Dermatol.* **2003**, *120*, 842–884. [CrossRef]

Review

Safety Testing of Cosmetic Products: Overview of Established Methods and New Approach Methodologies (NAMs)

Manon Barthe [1], Clarisse Bavoux [2], Francis Finot [3], Isabelle Mouche [3], Corina Cuceu-Petrenci [3], Andy Forreryd [4], Anna Chérouvrier Hansson [4], Henrik Johansson [4], Gregory F. Lemkine [5], Jean-Paul Thénot [1] and Hanan Osman-Ponchet [1],*

[1] PKDERM Laboratories, 45 Boulevard Marcel Pagnol, 06130 Grasse, France; Manon.barthe@pkderm.com (M.B.); jean-paul.thenot@pkderm.com (J.-P.T.)
[2] CEHTRA, 15 Rue Aristide Briand, 33150 Cenon, France; clarisse.bavoux@cehtra.com
[3] GenEvolutioN, 4 rue des Fréres Montgolfier, 78710 Rosny Sur Seine, France; francis.finot@genevolution.fr (F.F.); Isabelle.mouche@genevolution.fr (I.M.); corina.cuceu@genevolution.fr (C.C.-P.)
[4] SenzaGen AB, 22381 Lund, Sweden; andy.forreryd@senzagen.com (A.F.); Anna.C.Hansson@senzagen.com (A.C.H.); Henrik.Johansson@senzagen.com (H.J.)
[5] Laboratoire Watchfrog, 1 rue Pierre Fontaine, 91000 Evry, France; lemkine@watchfrog.fr
* Correspondence: hanan.osman.ponchet@pkderm.com

Abstract: Cosmetic products need to have a proven efficacy combined with a comprehensive toxicological assessment. Before the current Cosmetic regulation N°1223/2009, the 7th Amendment to the European Cosmetics Directive has banned animal testing for cosmetic products and for cosmetic ingredients in 2004 and 2009, respectively. An increasing number of alternatives to animal testing has been developed and validated for safety and efficacy testing of cosmetic products and cosmetic ingredients. For example, 2D cell culture models derived from human skin can be used to evaluate anti-inflammatory properties, or to predict skin sensitization potential; 3D human skin equivalent models are used to evaluate skin irritation potential; and excised human skin is used as the gold standard for the evaluation of dermal absorption. The aim of this manuscript is to give an overview of the main in vitro and ex vivo alternative models used in the safety testing of cosmetic products with a focus on regulatory requirements, genotoxicity potential, skin sensitization potential, skin and eye irritation, endocrine properties, and dermal absorption. Advantages and limitations of each model in safety testing of cosmetic products are discussed and novel technologies capable of addressing these limitations are presented.

Keywords: cosmetic product safety; non-animal-testing methodologies; dermal absorption; skin irritation; skin sensitization; genotoxicity; endocrine disruptors

1. Introduction

Cosmetic products need to have a proven efficacy combined with a comprehensive toxicological assessment. The 7th Amendment to the European Cosmetics Directive has banned animal testing for cosmetic products and for cosmetic ingredients in 2004 and 2009, respectively. Then, the European Cosmetic Regulation N°1223/2009 and the specific Regulation N°655/2013 specify the required data to proof the safety and support the claims. Largely driven by regulatory authorities, a wide range of alternatives to animal testing have been developed and validated for safety testing of cosmetic products and adopted as test guidelines (Figure 1). This review discusses the main in vitro alternative models used in safety testing of cosmetic products and cosmetic ingredients with a focus on regulatory requirements, genotoxicity potential, skin sensitization potential, skin and eye irritation, endocrine properties, and dermal absorption. Advantages and limitations of each model in safety testing of cosmetic products are discussed and novel technologies capable of addressing these limitations are presented.

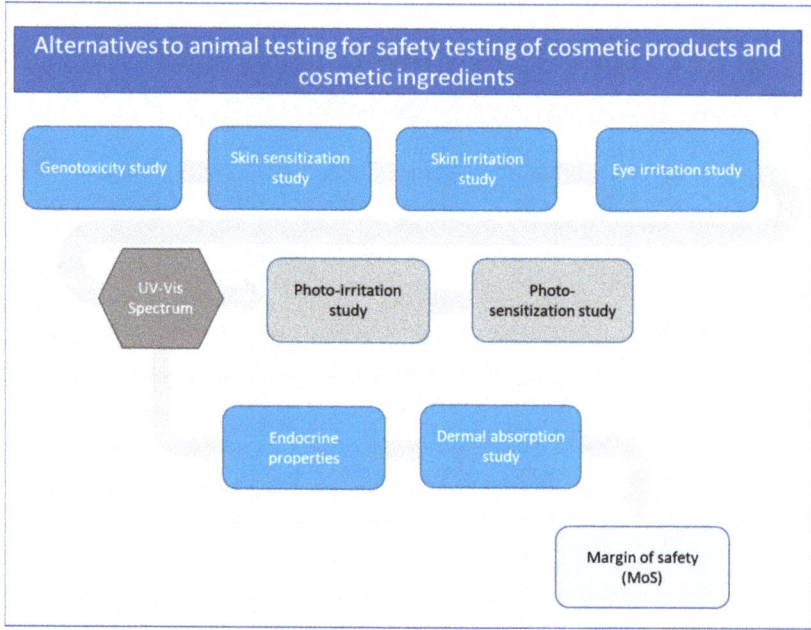

Figure 1. Overview of different alternatives to animal testing for safety assessment of cosmetic products and cosmetic ingredients. Assays in grey are not discussed in this review.

2. Regulatory Requirements for Cosmetics Safety Assessments

Overall Context

In Europe, the Cosmetic Regulation N°1223/2009 sets the framework for the safety of any cosmetic product [1]. Although, many other geographical areas do not specify the detailed documentation to establish their own frameworks, their regulations share the common goal of ensuring safety of the final consumers.

Some ingredients must be included in so-called "positive" lists, for the ones having specific functions (Annex VI for colorants, Annex V for preservatives, Annex V for UV filters). An ingredient with such a function should then comply to the requirements of the given Annex. Some ingredients are prohibited (Annex II) or restricted to particular uses (Annex III).

The origin of those regulatory limitations is mainly safety. In Europe, some of the ingredients are evaluated by the SCCS (Scientific Committee on Consumer Safety), which publishes its opinion with safe conditions of use, before the ingredient is listed in an annex. The SCCS publishes opinions based on the evidence presented to it, combined with guidance. That is helpful, rather than spelling out the prescriptive demand for strict adherence to precise regulatory "guidelines". The European committee regularly provides a guidance for the evaluation of the safety of ingredients [2,3]. In the USA, the CIR (Cosmetic Ingredient Review), established from a trade association (currently the PCPC) with the support of the FDA prioritizes and assesses cosmetic ingredients, generally consider groups of similar substances based on chemical families or plant-derived ingredients. The CIR's report does not include the risk assessment.

All regulated ingredients must have a favorable opinion of the SCCS, such as the recent ones on resorcinol, for its use in hair dyes [4], propylparaben as preservative (updated opinion discarding any concern related to endocrine disruption) [5] or octocrylene as UV filter (other update related to endocrine disruption) [6].

However, the committee can also give its opinion on substances for non-regulated uses (titanium dioxide in inhaled products [7] or aluminum in lipsticks [8]).

Some publications can also be available from national authorities, related to particular concern for a country (example of phenoxyethanol in France [9]), or specific investigations allowing a better management of the risk, as in the case of "technically unavoidable concentrations" of heavy metals, studied in Germany [10].

Transversal regulations can have consequences on the safety of the substances used in cosmetic products: the CLP Regulation (classification, labelling and packaging of substances and mixtures) [11] of major importance for CMR (carcinogenic, mutagenic and reprotoxic) substances. The carcinogenic, mutagenic, and reprotoxic substances are considered as the most dangerous substances; their harmonized classification in Europe is rarely based on epidemiological information (asbestos, benzene, etc.) and more generally based on experimental results in animals (musk xylene, Disperse Yellow 3, etc.). The Annex XVII of REACH can be of major importance for a very limited number of substances: D4 (cyclopentasiloxane) and D5 (cyclotetrasiloxane) are prohibited silicones in rinsed products above 0.1% (under entry 70 of the Annex XVII of REACH for restrictions) [12]. This decision is not triggered by toxicological properties but by their fate in the environment: these are the PBT and vPvB effects (for Persistent, Bioaccumulative and Toxic, Very Persistent, Very Bioaccumulative).

The list of SVHC (substances of very high concern) includes substances based on concern regarding reprotoxicity, carcinogenicity, endocrine disruption or effects for the environment, PBT or vPvB.

Those programs are somehow linked to each other (the general concern of endocrine disruption justified a call for data from the European Commission to revise the opinions of the SCCS (e.g., Benzophenone-3, octocrylene, benzyl salicylate ...) in the past two years.

- **Substances restricted by an Annex**

When the SCCS receives a mandate from the European Commission to assess the safety of a substance for a regulated function, the opinion is based on the analysis of the scientific dossier submitted by the industry.

The scientific opinion considers each endpoint, including local tolerance (skin irritation, phototoxicity when relevant), genotoxicity, systemic toxicity including reprotoxicity and sub-chronic/chronic toxicity. Characterization of dermal absorption is essential to calculate the SED (systemic exposure dose).

The exposure of the substance is considered as its expected concentration in cosmetic products, either in one given product or in several products, when a broad use is expected, as it would be for a preservative.

- **Substances not restricted by an Annex**

Any other substance, ingredient, or impurity has the obligation of being safe for the consumer, based on the toxicological profile, as required by the Annex I and Guidelines [13], using regularly updated data from supplier or literature.

There are two points of view: the one of the supplier of the ingredient and the one of the Responsible Person for a cosmetic product using the ingredient (the Responsible Person being the legal entity in Europe responsible for the product, generally the manufacturer). They do not have the same regulatory obligations. However, they should have the same purpose: consumer safety.

Any supplier of a cometic ingredient, such as any company which manufactures and markets a substance in the European Union, must register its substance according to the annual tonnage.

Even if the intrinsic toxicity of a substance is independent from its production, the number of toxicological results required in a REACH registration dossier depend on the annual tonnage. Highly toxic substances and substances of low toxicity have the same requirements (but important concern should be taken into account among the program of SVHC: substances of very high concern). No toxicological data are requested for substances registered below 1 to 10 tpa (ton per annuum) and increasing information is required to be submitted with increasing tonnage bands.

For tonnage of 10–100 tpa (Annex VII): toxicological requirements include data for in vitro skin irritation/corrosion, in vitro eye irritation, skin sensitization, in vitro gene mutation in bacteria, acute toxicity, and short-term toxicity (28 days).

At 10 to 1000 tpa (Annex VIII): toxicological requirements include data for in vitro mutagenicity study in mammalian cells or in vitro micronucleus study, in vitro gene mutation in mammalian cells, in vivo skin irritation, in vivo eye irritation, possibly testing proposal for in vivo genotoxicity, acute toxicity, and screening for reproductive/developmental toxicity.

At 100 to 1000 tpa (Annex IX) following endpoints are added: the sub-chronic toxicity (90 days), prenatal developmental toxicity in one species, and extended one-generation reproductive toxicity.

Finally, above 1000 tpa (Annex X) a long-term repeated dose toxicity (≥ 12 months) if triggered, with developmental toxicity in a second species, extended one-generation reproductive toxicity, and carcinogenicity.

Several reviews of these methods are available; we can cite a very recent one focused on cosmetic and REACH regulations [14]. Particularly, assessing the safety of the consumer should include the assessment of any potential regarding endocrine disruption, but this endpoint is not required in the REACH registration dossiers. The inclusion of such criteria by the CLP Regulation could change things in the future.

It is then important to realize that for ingredients produced below 10 to 1000 tpa, no information is available about the DNA damage (micronucleus test), and below 100 tpa, neither any sub-chronic toxicity nor any information on the full cycle of reproduction is known. A supplier of cosmetic ingredients should then think about the need of the cosmetic brands (Responsible Persons in general) who need to prove the safety of each ingredient.

The cosmetic brand (the Responsible Person) is the one responsible of the product. Studies can be made on the product, to confirm a good acceptability in humans. It is mostly to confirm the absence of eye and skin irritation, by in vitro test and other complementary tests in humans (the grail being the use test in normal conditions of use, to confirm the absence of objective irritation and absence of signs of discomfort). The tests for photo-toxicity or skin sensitization are rarely performed. It should be reminded that the Human Repeat Insult Patch Test (HRIPT) is non ethical and usually the historical data are significantly poor from a statistical point of view using a small size panel [15,16]. However, the new in vitro tests for skin sensitization are quite promising, particularly if they can cover multiple Key Events of the Adverse Outcome Pathway, and if they can be applied to the finished product. Both the SENS-IS and Genomic Allergen Rapid Detection (GARD) assays analyze the genomic response of the cells to the exposure of the substance or the product to predict sensitization, including its potency [17], with GARD assay being able to quantify the dose–effects relationship, thus providing a good perspective for its use in quantitative risk assessment [18]. Any test done on the finished product, as those two last ones, and the tests made on eleuthero-embryo from fish or amphibians discussed in this article are of particular relevance, since a large part of the risk assessment on the product in based on individual data of substances.

The major part of the safety then relies on the toxicological data of the substances. The toxicological results can come from the supplier, when they have a REACH registration dossier, or when they voluntary produce additional in vitro data. It can also be existing data from literature or in silico predictions Quantitative Structure-Activity Relationship (QSARs) or read-across. The safety assessor, working with the Responsible Person, makes a comprehensive search of existing toxicological information to write the toxicological profile of the ingredient, and possibly identify any data gaps. Pragmatically, toxicological profiles of ingredients often lack some information. Among the most current data gaps includes following endpoints: skin sensitization, DNA damage, chronic toxicity, and dermal absorption. With one exception, in vitro assays exist for all these endpoints, most of them with OECD guidelines, or with good results of validation. When it is chosen not to perform the test (data waiving), a rationale is absolutely needed as justification. In vitro micronucleus test is one of the missing test which has no reason to be lacking, since an in vitro OECD test exists for a long time. Probably there is a misunderstanding of the Responsible Person who might not realize that it is absolutely complementary to the in vitro mutagenicity test in bacteria, since both tests investigate two independent types of abnormalities of DNA, both predictive of cancer.

In some cases, a reliable in silico prediction, with one, or even better, consensus from several complementary software, can waive or replace such tests. This solution can be cheaper than testing and the rationale can be robust. In silico predictions are also a good strategy when associated to partially concluded results, such as the in vitro mutagenicity test. This test is not sufficient to investigate genotoxicity, but a QSAR prediction can provide a good orientation before performing the in vitro micronucleus assay, to better understand the potential of a substance to induce DNA damage. Such approaches are widely accepted for the regulatory assessment of pharmaceutical impurities under ICH M7 guideline [19].

Currently, with other methods gathered in the so-called NAMs (New Approach Methodologies), read-across is a major tool to predict the systemic toxicity of a substance in the absence of any animal testing. Finding structural analogues, selecting them based on relevant criteria, and predicting an endpoint-specific toxicity based on the results formerly obtained with those analogues is both a very ethical way to use existing data, and provides a relevant and reliable solution for predicting sub-chronic/chronic toxicity and reprotoxicity [20]. This parameter is one of the criteria of toxicokinetic (absorption, distribution, metabolism, and excretion; ADME) which should be better used in the future to enhance the application of NAMs [21].

Last but not least, although dermal absorption could help calculating a precise margin of safety, it is hardly investigated. This rare information is of equal importance in the calculation of the MoS (Margin of Safety) as the systemic NOAEL (or Point of Departure) and the exposure. Generally unknown, it is, by default, estimated to 50% according to the Notes of Guidance from the SCCS. For some substances, a "very low rate" can justify to avoid investigating systemic toxicity. The mathematical modeling of dermal absorption is an important field of research [22] but no robust model is currently available. Some models identified good predictivity but were limited to small substances below 300 Da [23]. A recent preliminary retrospective analysis of the ingredients with opinions of the SCCS showed that physicochemical properties of the substance can differentiate the ones with low and high dermal absorption (the threshold being at 2%) [24].

This article does not detail the requirement on impurities, which also deserve the attention of the safety assessor. CMR impurities are prohibited, but we can recommend to pay attention to any impurity, since this could have adverse effects.

3. Genotoxicity Assessment of Cosmetic Products

In the second part of the 20th century, many research teams [25] have developed different kind of tests based on different mechanisms showing direct DNA damages (DNA adduct, unscheduled DNA synthesis, DNA repair chromosomal aberrations), to detect direct DNA reactive substances that alter DNA and therefore the genetic code. In the 70s, Bruce Ames developed the most famous bacterial Reverse Mutation test, the "Ames test" [26]. The most relevant mutagen tests were quickly taken into account by regulatory authorities to identify genotoxic substances in cosmetics [27] and also by cosmetics companies for optimization of the methods and refined cosmetics ingredients [28]. Test battery strategies for genotoxicity evaluation have been issued by regulatory agencies and guidelines are published by OECD.

In the safety assessment of cosmetic ingredients, the assessment of genotoxic potential is crucial. The SCCS 10th Revision [2] recommended to use an in vitro battery of two tests. One test for the evaluation of the potential for mutagenicity: bacterial reverse mutation test (OECD 471) Ames test [29] and a second in vitro micronucleus test (OECD 487) [30] for the evaluation of chromosome damage (clastogen and aneuploidy). The combination of both tests allowed the detection of all relevant genotoxic carcinogens [31,32]. The test system should be exposed to the test item both in the absence and in the presence of a metabolic activation system (S9-fraction from the livers of rats treated with Aroclor 1254 or a combination of phenobarbital and β-naphthoflavone) [33].

The mutagenicity: bacterial reverse mutation test should be performed in the first instance, as the result could lead to an end of the project. The nature of test item has an impact on the method that should be used and consequently on the expected result. For pure compounds, if using the Ames test, the structure of the test item should be considered. Thus, depending on the nature of the test article the metabolic activation system should be adapted (SCCS/1532/14). For nanoparticles, a gene mutation test in mammalian cells (OECD 476), or mouse lymphoma assay (OECD 490) should be performed instead of the Ames test. For complex mixtures such as biological compounds or plant extracts the, presence of amino acid producing a feeder effect could be observed. In this case "treat and wash" method [34,35] could be used. The presence of flavonoids i.e., quercetin or kaempferol in plant extract could lead to increases in the number of the revertant colonies [36], in such case the quantification of this kind of substances in the plant extract is essential to explain the results obtained [37,38].

Before engaging into the second genetic toxicology test, an in-silico assessment (Quantitative Structure-Activity Relationship QSAR, DEREK, Multicase, or Compound Toxicity Profile) is useful to predict the clastogen potential of the pure chemical in accordance with the stringent quality criteria and the validation principles laid down by the OECD 487 [39]. In case of alert or when the prediction is out of domain, the micronucleus test should be performed following OECD 487 guideline. Recently, this technic has been refined in order to avoid a "false positive". The cell lines (V79, CHO and CHL) were consistently more susceptible to cytotoxicity and micronucleus induction than p53-competent cells and are therefore more susceptible to giving misleading positive results. These data suggest that a reduction in the frequency of misleading positive results can be achieved by careful selection of the mammalian cell type for genotoxicity testing [40].

One of the strengths of the cosmetics industry is the exclusive use of in vitro tests and consequently in vitro micronucleus has been also adapted to high-throughput technology, i.e., with only 10 milligrams, a micronucleus test is performed by flow cytometry [41] or using automated slide image analysis systems [42] and with double labelling telomere and centromere the distinction between aneugen and clastogen effect could be done [43,44].

When the results from both tests are clearly negative, the test item has no mutagenic potential. On the other hand, when the results from both tests are clearly positive, the test item is considered as being mutagen. In both cases further testing is not mandatory.

When one of the two tests gives a positive result, the test item is considered an in vitro mutagen. Further testing is required for excluding mutagen (clastogen) potential of the test item assessed.

Equivocal results for mixture plant extract can be obtained in particular in micronucleus test when excessive osmolarity, pH or excessive concentration leads to a high level of cytotoxicity [43,44]. In this case the toolbox for further evaluation in WoE (weight of evidence) approach is described in the SCCS recommendation:

"The comet assay [45] in mammalian cells or on 3D-reconstructed human skin [46] is a tool which can support a WoE approach in the case of a positive or equivocal gene mutation test in bacteria or mammalian gene mutation test."

This battery of tests leads to the identification of substances named initiators. They and their metabolites are DNA reactive carcinogens. In the theory of carcinogenesis, a second kind of substances are the promotors, they are non-genotoxic carcinogens. The SCCS/1602/18 (2018) recommends using the cell transformation assay (CTA) [47,48] as an alternative new test to in vivo carcinogenesis studies, to detect genotoxic and non-genotoxic carcinogens.

Progress in the knowledge of stem cells makes it is possible to propose new biological models to be closer to the in vivo exposure such organoid models [49] or for a screening approach such as the ToxTracker® model. The total blood is also a robust alternative, as it is easily available and extensively studied. In silico, and in the next future, AI (artificial intelligence), for analysis and prediction will be increasingly relevant, with the concept to build a "finger print of genotoxicity" as for drug in pharmaceutical companies.

4. Skin Sensitization Assessment of Cosmetic Products

Skin sensitizers are chemicals that have the intrinsic potential to induce a state of hypersensitivity in humans, that upon repeated topical exposure may result in the development of allergic contact dermatitis (ACD). Sensitization involves the activation of an adaptive immune response and the priming of immunological memory, and once acquired, it is often a chronical condition, and elicitation of clinical symptoms can only be prevented by avoiding exposure to the inducing agent (see for example [50] for an excellent review). Proactive identification and evaluation of skin sensitization potential is therefore of central importance for safety evaluation of chemicals and represents a key toxicological endpoint among regulatory authorities across multiple industries, and not least for cosmetics, where the intended route of exposure often is via dermal application [51].

Before a new cosmetic ingredient is placed on the European market, evaluation of its safety profile, including the assessment of skin sensitization hazards and potency is mandatory. Following the revision of Annex VII of the REACH regulation [52], as well as the transformation of the cosmetic directive into a regulation (EC1223/2009) [1], traditional animal models, such as the Guinea Pig based assays (GPMT or the Buehler test) [53] or the murine Local Lymph Node Assay (LLNA) [54], are no longer allowed to meet the information requirements for substances exclusively intended for use in cosmetic products. To this end, a plethora of New Approach Methods (NAMs), such as in chemico and in vitro methods, have been validated and incorporated into official test guidelines by the OECD, serving as viable replacements for animal studies. These methods are designed to target individual Key Events (KE) in the Adverse Outcome Pathway (AOP) for skin sensitization [55], which recapitulates the most important key mechanistic events that are required for the development of skin sensitization. Currently, three technical Test Guidelines (OECD TG 442 C, D and E) describe a total of seven such methods, including the KE1 based Direct Peptide Reactivity Assay (DPRA) and the Amino acid Derivative Reactivity Assay (ADRA) [56], the KE2 based assays KeratinoSens and LuSens [57], and the KE3 based assays h-CLAT, U-SENS, and the IL-8 Luc assay [58]. According to the current testing paradigm, these methods should not be considered as stand-alone assays, but rather in the context of a tiered testing strategy, a so-called defined approach (DA), where a fixed data integration procedure is used to arrive at a final classification, based on the readout from several NAMs. Currently, several DAs have been described for hazard identification of skin sensitizers, and their individual components, data integration procedures (DIPs), and performances have been summarized in [59]. Importantly, based on the empirical data from this publication, accuracies of the proposed DAs, ranging between 75.6% to 85.0%, were superior to that of the LLNA (74.2%) for predicting human skin sensitization hazard. In addition to the current OECD adopted assays, several alternative and innovative assays are in the process of being validated and adapted as official TGs [60], some showing predictive performances similar to the proposed DAs, also when considered as stand-alone assays [61]. Thus, skin sensitization testing is an ever-moving target, and to provide guidance to testing and safety evaluation to the cosmetic industry, the Scientific Committee on Consumer Safety (SCCS) publishes the "Notes of Guidance for the Testing of Cosmetic Ingredients and Their Safety Evaluation" [2], ensuring that testing can be performed in compliance with EU cosmetic legislations.

Despite the above-mentioned progress to replace animal experimentation, more work is still needed to address certain limitations with current NAM-based strategies. For example, it has been recognized that certain chemicals of interest to the cosmetic sector may be difficult to test in the conventional OECD validated assays [62]. Such limitations, as far as they have been identified, are described in individual TGs, and may include constraints with testing of hydrophobic ingredients, pre-pro haptens, and complex substances, including natural extracts where the ingredient of concern is often present in minute concentrations within a complex mixture. Novel state-of-the-art scientific methods currently in the OECD Test Guideline Program (TGP) and under evaluation for official TG adaption [60], such as the Genomic Allergen Rapid Detection (GARD) assay [63,64], which is based on the measurements of a biomarker signature of genes associated with immunologically relevant pathways to the sensitization process, have shown promise to address some of these limitations. For example, the GARD assay is compatible with a variety of different solvents that can be applied to increase bioavailability of a Test Item [65], and a protocol is also available for testing of solid materials, such as medical devices, using both polar and non-polar extraction vehicles in compliance with ISO-10993:12 [66]. Such findings may prove potentially useful also for cosmetic-related test items, such as UVCBs or natural extracts with limited solubility in conventional assay solvents, such as DMSO or water. Furthermore, several 3D-models based on reconstructed human epidermis (RHE) have also been developed to address some of the solubility limitations (reviewed in [62]). The majority of these assays have a clearly defined readout of established biomarkers (e.g., IL-18), while others are less transparent. In a recent publication evaluating the performance of a selection of such models, the majority of the RHE-based assays showed similar, or slightly improved performances (dependent on the specific RHE-assay) to the best performing OECD validated assay, the h-CLAT assay, when investigating a limited set of "difficult-to-test" substances in comparison to human reference data [63], demonstrating that such assays may comprise a viable source of information within a weight-of- evidence approach for testing within this chemical domain.

In addition to the limited applicability domain, the most obvious limitation of the current OECD validated assays is likely that they have only been validated for skin sensitization hazard identification, and not for assessment of sensitizing potency, which is a critical component for risk assessment of cosmetic ingredients when used in consumer products. Skin sensitization is a threshold phenomenon, and a quantitative risk assessment (QRA) of individual ingredients aims to define a maximum dose of the chemical not inducing sensitization (referred to as the NESIL value) [67,68]. The general procedure for QRA, involving a continuous prediction of skin sensitizing potency as a point of departure (POD), which is subsequently adjusted by applying uncertainty factors, has been described for fragrances [67], and its applicability to general cosmetic ingredients is currently being discussed. Development of NAM based strategies also for continuous assessment of skin sensitizing potency for use as point-of-departure in the QRA is ongoing, and examples include the DA-based Artificial Neural Network Model for Predicting LLNA EC3 [69], as well as the recently proposed GARDskin Dose-Response model [18,70].

Finally, as novel NAM-based methods are developed to replace traditional animal models for assessment of cosmetic ingredients, the ultimate arbiter of the capacity of these tests to protect human health must be evaluated based on how well they correlate with reliable information on the skin sensitizing activity of chemicals in humans, and not how well they recapitulate the weaknesses of the "gold" standard animal tests, irrespective of their historical consideration as valid and adapted OECD methods. For chemicals of hitherto unknown sensitization potential, the preclinical evaluation of cosmetic ingredients using the NAM strategies described above is an essential and important first step to ensure the safety profile of cosmetics, but also as described in [71], post-market surveillance, often referred to as cosmetovigilance, will remain an important part to ensure that the use of cosmetic ingredients, as well as their concentration in formulated products remain safe to the consumers.

5. Endocrine Properties Assessment of Cosmetic Products

On the 13 December 2017 the European Parliament adopted scientific criteria to define endocrine disruptors which came into force for plant protection products and biocides in 2018 [72]. This has been a major step towards the future implementation of similar criteria for regulation of cosmetics in Europe. Despite the discrepancies due to the particular context of cosmetics, a few lessons relating to endocrine assessment strategies have been learnt from experience.

Adopted criteria for endocrine disruptors are closely related to the WHO definition of 2012 [73]. An endocrine disruptor is defined by three main criteria: its endocrine mode of action, its capacity to cause an adverse effect, and the plausible link between this endocrine activity and the related adverse outcome.

Regulatory authorities require datasets to permit a conclusive assessment on the disruptive capacity of an endocrine active sample. However, for cosmetic ingredients this will be difficult as availability of comprehensive endocrine test systems is very limited without accessing animal experimentation. Therefore, alternative models will be required to overcome this difficulty that can provide data which will contribute to safety of cosmetics for the endocrine system in an ethical manner.

Since 2002, experts representing OECD member countries have published test guidelines dedicated to endocrine assessment of chemicals. These internationally acknowledged methods are listed, and their proper usage is described within the OECD Guidance Document 150 [74]. According to this document, adversity should be assessed (using laboratory animals) to achieve a conclusive assessment of an endocrine disruptor. OECD validated methods cover so far EATS (Estrogen, Androgen, Thyroid, and Steroidogenic) endocrine pathways, for which specific adverse physiological outcomes have been characterized.

It could be argued that the absence of endocrine activity excludes the need for investigating physiological adversity. This opens a possible testing strategy for an ethical cosmetic approach: using a battery of validated in vitro/embryonic models to cover all major modes of action of endocrine disruptors on EATS pathways. Cellular-based assays using tumoral cell lines, allow the assessment of the transactivation capacity estrogen (OECD TG 455) [75] or androgen (TG 458) [76] receptors, as well as disruption of steroidogenesis (TG 456) [77]. Nevertheless, performing all these assays independently will not mimic the interaction of these mechanisms occurring in vivo and many modes of actions are not covered by in vitro tests such as disruption of 5-alpha reductase endocrine target to counteract alopecia [78]. The complexity and crosstalk of endocrine pathways as well as the number of mechanisms involved often leads to false positive or false negative results using cellular models [79,80]. Identifying an endocrine disruptor boils down to elucidating an adverse outcome pathway and requires a complete endocrine system as a model.

As indicated by the SCCS guidance notes [3], due to the conservation of endocrine mechanisms across vertebrate species data provided by "some ecotox tests may be informative for the assessment of the endocrine activity of a compound in humans". This is of great value as the additional information provided by ecotoxicological tests significantly increases the weight of evidence available for endocrine assessment of cosmetic ingredients.

Embryos of aquatic vertebrates provide ethical and useful models to assess endocrine activity of cosmetic ingredients or products in a whole endocrine system. In 2019, the OECD published the first eleuthero–embryo-based test to assess Thyroid activity, Test Guideline 248 (XETA) [81]. Eleuthero–embryo defines early life stages post-hatch which still depend on maternally deposited energy reserves making them eligible for cosmetic testing according to the EU definition of a laboratory animal [82]. This first eleuthero–embryonic model for measuring thyroid activity paved the way for the development of a series of embryonic models derived from fish and amphibians bearing fluorescent reporter constructs integrating hormonal responsive elements.

Among assays in the OECD process of validation, the EASZY and REACTIV assays are dedicated to measuring estrogenic activities. These models carry specific targets to reveal the brains response to estrogens (EASZY) [83] and estrogenic control over reproduction (REACTIV) [84]. Further, it is also included in the OECD work program on endocrine disruptors and in the EFSA/ECHA guidance document [85] on endocrine disruptor assessment is the RADAR [86] assay which measures androgenic activities related to male reproductive behaviors.

These embryonic models allow the detection and quantification of endocrine activities by the quantification of fluorescence. Even if these in vitro aquatic models are not necessarily predictive of the effects in humans, they make it possible to detect endocrine activity and constitute a predictive screening tool.

The criteria adopted by the EU for the assessment of endocrine disruptors are hazard based. These criteria were implemented within plant protection product and biocide regulations in 2018. Weight of evidence provided by models that identify modes of action and related adverse outcomes have replaced risk assessment for the classification of endocrine disruptors. However, for the assessment of cosmetic ingredients, implementation of these hazard-based criteria without the use of laboratory animals remains a challenge. Despite this, some solutions are available to provide more realistic exposure scenarios whilst avoiding the use of regulated life stages of laboratory animals. Linking the selection of test concentrations for hazard assessment to a range of daily doses of a compound or product could be one approach for screening cosmetics. Recent advances in the development of eleuthero–embryonic tests systems also provide options for semi-quantitative assessment of endocrine activity in a whole endocrine system. Allowing the identification of ingredients, extracts, or preparations, which would require more in-depth investigation.

Data provided by embryonic models and cellular assays will be a great source of knowledge to feed into the development of in silico models. Our ultimate aim should be to develop in silico models of each endocrine pathway, and one day perhaps a computational model of a complete vertebrate endocrine system.

6. Assessment of Dermal Absorption of Cosmetic Products

Assessment of dermal absorption is a crucial aspect of cosmetic product and ingredient safety, as opposed to drugs, which almost always enter the body in other ways. In vitro dermal absorption studies are the gold standard method for skin pharmacokinetic evaluation and are suitable to predict the expected dermal absorption by humans.

The purpose of the dermal absorption testing, also known as dermal penetration or percutaneous penetration, is to provide a measurement of the absorption or penetration of a substance through the skin barrier and into the skin.

Detailed guidance on the performance of in vitro skin absorption studies is available (OECD 2004, 2011, 2019), [87–89]. In addition, the SCCNFP (Scientific Committee on Cosmetics and NonFood Products) adopted a first set of "Basic Criteria" for the in vitro assessment of dermal absorption of cosmetic ingredients back in 1999 and updated in 2003 (SCCNFP/0750/03) [90]. The SCCS updated this Opinion in 2010 (SCCS/1358/10) [91]. A combination of OECD 428 guideline with the SCCS "Basic Criteria" (SCCS/1358/10) is considered to be essential for performing appropriate in vitro dermal absorption studies for cosmetic ingredients.

Dermal absorption studies are conducted to determine how much of a chemical penetrates the skin, and thereby whether it has the potential to be absorbed into the systemic circulation. Therefore, knowledge of dermal absorption phenomena is essential for:

- Safety issues: the presence of systemic test item may lead to systemic adverse effects, the quantities absorbed is taken into consideration in toxicological risk assessment to extrapolate human exposure and calculate the margin of safety (MoS); and
- Therapeutic aspects: the quantities penetrated can be taken into consideration to predict the therapeutic concentration at the target sites in skin tissue.

In vitro dermal absorption studies are applied in different sectors and for different purposes:

- Formulation Screening: for selection of lead candidate formulation;
- Bioequivalence: to determine if the new product has the same degree of dermal absorption as reference product. In vitro dermal absorption assay was recently used to demonstrate bioequivalence, and the results of the comparison were accepted by the FDA in connection with the marketing authorization for Lotrimin Ultra cream [92];
- Cosmetics and consumer products: Dermal absorption rate is part of the toxicological profile of any ingredient. Almost always provided for any submission to the SCCS,

the in vitro dermal absorption studies can then be part of the safety assessment of a cosmetic product;
- Pharmaceutical products: in vitro dermal absorption studies are part of safety and efficacy assessment of topical products;
- Chemical/agrochemical: in vitro dermal absorption studies are part of safety assessment purposes. With respect to pesticides, the results of the in vitro dermal absorption studies alone are accepted for pesticides risk assessment purposes in the European Union and other countries.

Different types of formulations can be assessed through in vitro dermal absorption studies: creams, gels, ointments, suspensions, foam, patches, aqueous, solvent, hair dyes, shampoo, foundation, moisturizer, cleansers, soaps, sunscreen, etc.

When conducting in vitro dermal absorption study, skin sample is placed between two chambers (a donor chamber and a receptor chamber) of a Franz-type diffusion cell in a way such that the *stratum corneum* is facing the donor compartment where the formulation to be examined is applied, while the dermis is touching receptor compartment.

Human skin samples are usually obtained from patients undergoing plastic surgery. Abdominal skin is most convenient, due to the large areas that may be available. Carefully handled frozen human skin are suitable for testing the passive permeation of chemicals, when skin viability and metabolic activity were not being investigated [93]. However, for studies requiring the presence of viable epidermal tissue, such as investigations of drug transporters [94–98] or skin metabolism [96], fresh skin samples are required.

There are considerable differences in skin absorption across different body sites, attributed to *stratum corneum* thickness, hydration, and lipid composition [99–103]. To reduce variability, it is recommended to use split-thickness skin. Full-thickness skin is cut to approximately 500–750 µm using a dermatome. Quality of skin samples have to be checked at the beginning of the experiment. This is done by measuring transepidermal water loss (TEWL) indicative of barrier integrity.

A finite dose of tested product is applied on the skin surface and incubation is done at 32 °C. The permeation rate of a test item from the donor compartment through the skin into the receptor is determined by measuring the amount of drug in skin samples and in receptor fluid. Different analytical methods can be used to quantify the amount of test item in the samples.

Different analytical methods can be used to quantify concentration of test substance in different skin compartments according to physicochemical properties of the test substance such as lipophilicity, molecular weight, charge, and concentration of the test substance: liquid chromatography–tandem mass spectrometry (LC-MS/MS); inductively coupled plasma–tandem mass spectrometry (ICP-MS/MS), liquid chromatography with UV detection (LC-UV) or fluorescence detection (LC-Fluo), liquid scintillation counting (LSC) for radiolabelled compound, and imaging approaches, e.g., epifluorescence or confocal microscopy in the case of fluorescent molecules or matrix-assisted laser desorption–mass spectrometry imaging (MALDI-MSI) [104].

In vitro dermal absorption assay is very operator-dependent, and care needs to be taken especially when handling skin samples and when removing the excess of formulation. The success of the assay is equally dependent on the development and validation of sensitive analytical methods to quantify the amount of test substance in the samples.

One of the main challenges is how to measure dermal absorption in babies and infant skin necessary in cosmetic ingredient safety assessments. It is recognized that babies, infants, and children represent a distinct subpopulation for risk and safety assessments, and routinely considered the greater skin–surface area to body–mass ratio in children when performing cosmetic ingredient safety assessments [105]. Systemic exposures in babies and infants are generally assumed to be greater than in older children and adults. On one side, the percutaneous absorption could be higher because of the immaturity of the skin as a barrier to absorption (higher pH of the skin yields decreased barrier function and increased risk of irritation), particularly onto the nappy area. On the other side, the greater body–surface-area to body–mass ratio of babies and infants compared with older children and adults, mathematically induces high amounts in mg/kg bw/w for a similar quantity of product [106–110]. Modifications of existing in vitro skin penetration protocols to evaluate the potential for higher absorption from topically applied products are needed. The use of compromised skin represents a good alternative to mimic underdeveloped barrier function as in premature infant skin. Compromised skin can be achieved by different procedures, e.g., tape stripping, microneedling device, abrasive skin preparation pad, or even iontophoresis [111–113].

7. Skin and Eye Irritation Assessment of Cosmetic Products

Assessment of skin and eye irritation potential of an ingredient or formulation is an important part in cosmetic ingredient safety assessments.

Dermal irritation is defined as the production of reversible damage of the skin, following the application of a test substance for up to 4 h (OECD 404) [114]. Eye irritation is defined as the occurrence of changes in the eye following the application of a test substance to the anterior surface of the eye, which are fully reversible within 21 days of application (OECD 405) [115].

Skin and eye irritation are assessed using reconstructed human tissue-based test methods. Commercially available 3D-models based on reconstructed human epidermis (RhE) are used for skin irritation testing (OECD test Method 439) [116] and 3D-model based on reconstructed human cornea-like epithelium (RhCE) is used for eye irritation testing (OECD Test Method 492) [117]. It should be noted that there are different in vitro models that address serious eye damage and/or identification of chemicals not triggering classification for eye irritation or serious eye damage [3], but we will only focus on RhCE model.

The overall design 3D-models based on reconstructed human tissues mimics the biochemical and physiological properties of the upper layers of the human skin and eye.

RHE is a skin model composed of living human keratinocytes which have been cultured to form a multi-layered, highly differentiated epidermis. The model consists of highly organized basal cells and includes a functional skin barrier with an in vivo-like lipid profile.

RhCE is a corneal model composed of living human cells which have been cultured to form a multi-layered, differentiated corneal epithelium. The model consists of highly organized basal cells which progressively flatten out as the apical surface of the tissue is approached, analogous to the normal human in vivo corneal epithelium.

In both models, the cells are both metabolically and mitotically active, and release many of the pro-inflammatory agents (cytokines) known to be important in irritation and inflammation. Reconstructed human tissues are grown on special platforms at the air-liquid interface.

The test item is applied directly to the tissue surface, providing a good model of "real life" exposure. The endpoint used in both RhE and RhCE test methods is the cell-mediated reduction of MTT (3-(4,5)-dimethyl-2-thiazolyl-2,5-dimethyl-2H-tetrazolium bromide) into a blue formazan salt that is quantitatively measured after extraction from the tissues. A second endpoint can be used to increase sensitivity is the measurement of interleukin-1α (IL-1α) production.

If the viability is greater than 50% (RhE) or 60% (RhCE), the test item is classified as Non-Irritant (no-label or UN GHS No Category).

If the viability is below or equal to 50% in the case of RhE model, the test item is classified Irritant (UN GHS Category 2).

If the viability is below or equal to 60% in the case of RhCE, no prediction can be made, and further testing may be required.

So far, neither a single in vitro assay nor a testing battery has been validated as a standalone replacement for the in vivo test. New test systems are under development using stem cells. These could generate new alternatives for in vitro ocular toxicity testing [118].

8. Conclusions

The total number of experiments in animals only slightly decreased in Europe between 2015 and 2017. It changed from 9.59 million animals to 9.39 million, when it was 11.5 million in 2011. Animals are mainly used for research (69%) and then for regulatory purpose (23%). In 2017, 61% of the experiments in animals were for medical products for humans, 15% for veterinary products, 11% for industrial chemicals. Moreover, the report of the European Commission identifies a concern about the use of animals for endpoints where alternative methods exist (irritation, skin sensitization).

Despite the marketing ban of cosmetic ingredients and cosmetic products tested in animals, there is still debate on this issue. From a regulatory point of view, the position of the European Agency is clear and has been clarified ("Clarity on interface between REACH and the Cosmetics Regulation"). No cosmetic product is currently tested in animals in Europe. The cosmetic ingredients can have former results obtained from toxicological tests in animals. These results can be obtained after the animal testing ban, but only if required by another regulation (food, pharmaceutical, or even REACH, considering the obligations of safety of the workers). If cosmetics are the only use of a substance, all in silico and in vitro tests will then be encouraged to demonstrate the safety. However, for a toxicologist, it remains a huge challenge to guarantee the absence of risk based on the current available methods. All so-called New Approach Methodologies, using AOPs, IATAs, or Defined Approaches will be the foundation of the safety for future new ingredients [119].

A wide range of in vitro models for safety testing of cosmetic products and cosmetic ingredients has been developed and adopted in test guidelines. There is still an increasing need, largely driven by regulatory authorities and industry, to develop in vitro models to predict carcinogenicity, repeat dose toxicity and reproductive toxicity, for which no alternative in vitro methods are currently available.

Author Contributions: Conceptualization, H.O.-P., C.B., F.F., A.F., A.C.H., G.F.L. and J.-P.T.; validation, H.J., M.B. and I.M.; data curation, M.B., H.J., I.M., C.C.-P., A.F., C.B. and G.F.L.; writing–original draft preparation, M.B., C.B., F.F., A.F., A.C.H., H.J., G.F.L. and H.O.-P.; writing–review and editing, M.B., C.B., A.F., F.F., G.F.L. and H.O.-P.; supervision, J.-P.T. All authors have read and agreed to the published version of the manuscript.

Funding: This research received no external funding.

Institutional Review Board Statement: Not applicable.

Informed Consent Statement: Not applicable.

Data Availability Statement: There is no data supporting reported results generated during the study.

Acknowledgments: The authors thank Faizan Sahigara for assistance with the language in this article.

Conflicts of Interest: The authors declare no conflict of interest in this work.

References

1. European Parliament. Regulation (EC) No 1223/2009 of the European Parliament and of the Council of 30 November 2009 on Cosmetic Products. *Off. J. Eur. Union* **2009**, *L396*, 1–1355. Available online: http://data.europa.eu/eli/reg/2009/1223/oj (accessed on 7 June 2021).
2. SCCS (Scientific Committee on Consumer Safety). SCCS Notes of Guidance for the Testing of Cosmetic Ingredients and Their Safety Evaluation 10th Revision, 24–25 October 2018, SCCS/1602/18. Available online: https://ec.europa.eu/health/sites/default/files/scientific_committees/consumer_safety/docs/sccs_o_224.pdf (accessed on 7 June 2021).
3. SCCS (Scientific Committee on Consumer Safety). SCCS Notes of Guidance for the Testing of Cosmetic Ingredients and Their Safety Evaluation 11th Revision, 30–31 March 2021, SCCS/1628/21. Available online: https://ec.europa.eu/health/sites/default/files/scientific_committees/consumer_safety/docs/sccs_o_250.pdf (accessed on 7 June 2021).
4. SCCS (Scientific Committee on Consumer Safety). Opinion on Resorcinol (CAS No 108-46-3, EC No 203-585-2), Preliminary Version of 16 October 2020, Final Version of 30–31 March 2021, SCCS/1619/20. Available online: https://ec.europa.eu/health/sites/default/files/scientific_committees/consumer_safety/docs/sccs_o_241.pdf (accessed on 7 June 2021).

5. SCCS (Scientific Committee on Consumer Safety). Opinion on Propylparaben (CAS No 94-13-3, EC No 202-307-7), Preliminary Version of 27–28 October 2020, Final Version of 30–31 March 2021, SCCS/1623/20. Available online: https://ec.europa.eu/health/sites/default/files/scientific_committees/consumer_safety/docs/sccs_o_243.pdf (accessed on 7 June 2021).
6. SCCS (Scientific Committee on Consumer Safety). Opinion on Octocrylene (CAS No 6197-30-4, EC No 228-250-8), Preliminary Version of 15 January 2021, Final Version of 30–31 March 2021, SCCS/1627/21. Available online: https://ec.europa.eu/health/sites/default/files/scientific_committees/consumer_safety/docs/sccs_o_249.pdf (accessed on 7 June 2021).
7. SCCS (Scientific Committee on Consumer Safety). Opinion on Titanium Dioxide (TiO2), Preliminary Version of 7 August 2020, Final Version of 6 October 2020, SCCS/1617/20. Available online: https://ec.europa.eu/health/sites/default/files/scientific_committees/consumer_safety/docs/sccs_o_238.pdf (accessed on 7 June 2021).
8. SCCS (Scientific Committee on Consumer Safety). Addendum to the Scientific Opinion SCCS/1613/19 on the Safety of Aluminium in Cosmetic Products (Lipstick); SCCS/1626/20. Available online: https://ec.europa.eu/health/sites/default/files/scientific_committees/consumer_safety/docs/sccs_o_248.pdf (accessed on 7 June 2021).
9. ANSM. Evaluation du Risque Lié à L'utilisation du Phénoxyéthanol Dans Les Produits Cosmétiques. Available online: http://dev4-afssaps-marche2017.integra.fr/var/ansm_site/storage/original/application/58033db1a0bd86f6df50cf80b03e1839.pdf (accessed on 7 December 2016).
10. BVL (Bundesamt für Verbraucherschutz und Lebensmittelsicherheit). Technically avoidable heavy metal contents in cosmetic products. *J. Consum. Prot. Food Saf.* **2017**, *12*, 51–53. [CrossRef]
11. European Parliament. Regulation (EC) 1272/2008 of the European Parliament and of the Council. *Off. J. Eur. Union* **2008**, *L 353*, 1–1355.
12. ECHA. *ANNEX XVII TO REACH–Conditions of Restriction. Entry 70. Octamethylcyclotetrasiloxane (D4) Mdi*; ECHA: 2010. Available online: https://echa.europa.eu/documents/10162/50e79685-efaf-ac9a-4acb-d8be3f0e9ddc (accessed on 7 June 2021).
13. European Commission. *Commission Implementation Decision of 25 November 2013 on Guidelines on Annex I to Regulation (EC) No 1223/2009 of the European Parliament and of the Council on Cosmetic Products*; OJEU 2013/674/UE.; European Commission: Brussels, Belgium, 2013; Available online: http://data.europa.eu/eli/dec_impl/2013/674/oj (accessed on 7 June 2021).
14. Pistollato, F.; Madia, F.; Corvi, R.; Munn, S.; Grignard, E.; Paini, A.; Worth, A.; Bal-Price, A.; Prieto, P.; Casati, S.; et al. Current EU regulatory requirements for the assessment of chemicals and cosmetic products: Challenges and opportunities for introducing new approach methodologies. *Arch. Toxicol.* **2021**, *1*, 3. [CrossRef]
15. SCCS (Scientific Committee on Consumer Safety). Memorandum on Use of Human Data in Risk Assessment of Skin Sensitisation, SCCS/1567/15, 15 December 2015. Available online: https://ec.europa.eu/health/scientific_committees/consumer_safety/docs/sccs_s_010.pdf (accessed on 7 June 2021).
16. AFSSAPS. *Test Clinique Final de Sécurité D'un Produit Cosmétique en Vue de Confirmer Son Absence de Potentiel Sensibilisant Cutané Retardé: Recommandations*; AFSSAPS: 2008. Available online: https://ansm.sante.fr/documents/reference/recommandations-pour-les-produits-cosmetiques (accessed on 7 June 2021).
17. Gilmour, N.; Kern, P.S.; Alépée, N.; Boislève, F.; Bury, D.; Clouet, E.; Hirota, M.; Hoffmann, S.; Kühnl, J.; Lalko, J.F.; et al. Development of a next generation risk assessment framework for the evaluation of skin sensitisation of cosmetic ingredients. *Regul. Toxicol. Pharmacol.* **2020**, *116*, 104721. [CrossRef] [PubMed]
18. Johansson, H.; Gradin, R.; Forreryd, A.; Schmidt, J.; Id, A. Poster SOT: Quantitative Sensitizing Potency Assessment Using GARD Skin; 2021. Available online: https://senzagen.com/2021/04/15/poster-presented-at-sot-2021-quantitative-sensitizing-potency-assessment-using-gardskin-dose-response/ (accessed on 7 June 2021).
19. Fioravanzo, E.; Bassan, A.; Pavan, M.; Mostrag-Szlichtyng, A.; Worth, A.P. Role of in silico genotoxicity tools in the regulatory assessment of pharmaceutical impurities. *SAR QSAR Environ. Res.* **2012**, *23*, 257–277. [CrossRef] [PubMed]
20. Rovida, C.; Barton-Maclaren, T.; Benfenati, E.; Caloni, F.; Chandrasekera, P.C.; Chesné, C.; Cronin, M.T.D.; De Knecht, J.; Dietrich, D.R.; Escher, S.E.; et al. Internationalization of read-across as a validated new approach method (NAM) for regulatory toxicology. *ALTEX* **2020**, *37*, 579–606. [CrossRef]
21. Rogiers, V.; Benfenati, E.; Bernauer, U.; Bodin, L.; Carmichael, P.; Chaudhry, Q.; Coenraads, P.J.; Cronin, M.T.D.; Dent, M.; Dusinska, M.; et al. The way forward for assessing the human health safety of cosmetics in the EU-Workshop proceedings. *Toxicology* **2020**, *436*, 152421. [CrossRef] [PubMed]
22. Tsakovska, I.; Pajeva, I.; Al Sharif, M.; Alov, P.; Fioravanzo, E.; Kovarich, S.; Worth, A.P.; Richarz, A.N.; Yang, C.; Mostrag-Szlichtyng, A.; et al. Quantitative structure-skin permeability relationships. *Toxicology* **2017**, *387*, 27–42. [CrossRef]
23. Shen, J.; Kromidas, L.; Schultz, T.; Bhatia, S. An in silico skin absorption model for fragrance materials. *Food Chem. Toxicol.* **2014**, *74*, 164–176. [CrossRef]
24. Ates, G.; Steinmetz, F.P.; Doktorova, T.Y.; Madden, J.C.; Rogiers, V. Linking existing in vitro dermal absorption data to physico-chemical properties: Contribution to the design of a weight-of-evidence approach for the safety evaluation of cosmetic ingredients with low dermal bioavailability. *Regul. Toxicol. Pharmacol.* **2016**, *76*, 74–78. [CrossRef]
25. Kilbey, B.J.; Legator, M.; Nichols, W.; Ramel, C. (Eds.) *Handbook of Mutagenesis Test Procedures 1973 and Novel Edition*; Elsevier: Amsterdam, The Netherlands, 1984.
26. Ames, B.N.; Mccann, J.; Yamasaki, E. Methods for detecting carcinogens and mutagens with the Salmonella/mammalian-microsome mutagenicity test. *Mutat. Res.* **1975**, *31*, 347–364. [CrossRef]

27. Marzin, D. La mutagénèse, principes, méthodes d'étude et législation. Parfums Cosmétiques Arôme N°32. In Proceedings of the Conférence Présentée Devant La SFC Paris, Paris, France, 25 October 1979.
28. Shahin, M.M.; Chopy, C.; Mayet, M.J.; Lequesne, N. Mutagenicity of structurally related aromatic amines in the Salmonella/mammalian microsome test with various S-9 fractions. *Food Chem. Toxicol.* **1983**, *21*, 615–619. [CrossRef]
29. OECD. (1997-corrected 26 June 2020). *Test No. 471. Bacterial Reverse Mutation Test, OECD Guideline for the Testing of Chemicals*. Available online: https://www.oecd-ilibrary.org/docserver/9789264071247-en.pdf?expires=1623341347&id=id&accname=guest&checksum=03521B29829FC1D20763B99E2FF5DA3B (accessed on 7 June 2021).
30. OECD. *Test No. 487: In Vitro Mammalian Cell Micronucleus Test, OECD Guidelines for the Testing of Chemicals*; Section 4; OECD: Paris, France, 2016. [CrossRef]
31. Kirkland, D.; Aardema, M.; Müller, L.; Makoto, H. Evaluation of the ability of a battery of three in vitro genotoxicity tests to discriminate rodent carcinogens and non-carcinogens II. Further analysis of mammalian cell results, relative predictivity and tumour profiles. *Mutat. Res.* **2006**, *608*, 29–42. [CrossRef] [PubMed]
32. Kirkland, D.; Reeve, L.; Gatehouse, D.; Vanparys, P. A core in vitro genotoxicity battery comprising the Ames test plus the in vitro micronucleus test is sufficient to detect rodent carcinogens and in vivo genotoxins. *Mutat. Res.* **2011**, *721*, 27–73. [CrossRef]
33. Maron, D.M.; Ames, B.N. Revised methods for the Salmonella mutagenicity test. *Mutat. Res.* **1983**, *113*, 173–215. [CrossRef]
34. Brusick, D. Genotoxic effects in cultured mammalian cells produced by low pH treatment conditions and increased ion concentrations. *Environ. Mutagenesis* **1986**, *8*, 879–886. [CrossRef]
35. Aardema, M.J.; Galloway, S.; Zeiger, E.; Cimino, M.C.; Hayashi, M. Guidance for understanding solubility as a limiting factor for selecting the upper test concentration in the OECD in vitro Micronucleus Assay Test Guideline No. 487. *Mutat. Res.* **2011**, *722*, 89–90. [CrossRef]
36. Thompson, C.; Morley, P.; Kirkland, D.; Proudlock, R. Modified bacterial mutation test procedures for evaluation of peptides and amino acid-containing material. *Mutagenesis* **2005**, *20*, 345–350. [CrossRef]
37. Hamel, A.; Roy, M.; Proudlock, R. *The Bacterial Reverse Mutation Test. Genetic Toxicology Testing*; Elsevier Inc.: Amsterdam, The Netherlands, 2016. [CrossRef]
38. Schimmer, O.; Häfele, F.; Krüger, A. The mutagenic potencies of plant extracts containing quercetin in Salmonella typhimurium TA98 and TA100. *Mutat. Res.* **1988**, *206*, 201–208. [CrossRef]
39. OECD. Principles for the Validation, for Regulatory Purposes, of (Quantitative) Structure-Activity Relationship Models. Available online: http://www.oecd.org/chemicalsafety/risk-assessment/37849783.pdf (accessed on 21 May 2021).
40. Fowler, P.; Smith, K.; Young, J.; Jeffrey, L.; Kirkland, D.; Pfuhler, S.; Carmichael, P. Reduction of misleading ("false") positive results in mammalian cell genotoxicity assays. I. Choice of cell type. *Mutat. Res.* **2012**, *742*, 11–25. [CrossRef] [PubMed]
41. Bryce, S.M.; Bemis, J.C.; Avlasevich, S.L.; Dertinger, S.D. In vitro micronucleus assay scored by flow cytometry provides a comprehensive evaluation of cytogenetic damage and cytotoxicity. *Mutat. Res.* **2012**, *630*, 78–91. [CrossRef]
42. Verma, J.R.; Rees, B.J.; Wilde, E.C.; Thornton, C.A.; Jenkins, G.J.S.; Doak, S.H.; Johnson, G.E. Evaluation of the automated MicroFlow® and Metafer™ platforms for high-throughput micronucleus scoring and dose response analysis in human lymphoblastoid TK6 cells. *Arch. Toxicol.* **2017**, *91*, 2689–2698. [CrossRef] [PubMed]
43. M'Kacher, R.; Maalouf, E.E.; Ricoul, M.; Heidingsfelder, L.; Laplagne, E.; Cuceu, C.; Hempel, W.M.; Colicchio, B.; Dieterlen, A.; Sabatier, L. New tool for biological dosimetry: Reevaluation and automation of the gold standard method following telomere and centromere staining. *Mutat. Res.* **2014**, *770*, 45–53. [CrossRef] [PubMed]
44. Finot, F.; Kaddour, A.; Morat, L.; Mouche, I.; Zaguia, N.; Cuceu, C.; Souverville, D.; Négrault, S.; Cariou, O.; Essahli, A.; et al. Genotoxic risk of ethyl-paraben could be related to telomere shortening. *J. Appl. Toxicol.* **2017**, *37*, 758–771. [CrossRef]
45. Chetelat, A.A.; Albertini, S.; Gocke, E. The Photomutagenicity of fluoroquinolones in tests for gene mutation, chromosomal aberration, gene conversion and DNA breakage (Comet assay). *Mutagenesis* **1996**, *11*, 497–504. [CrossRef]
46. Barcham, R.; Orsini, N.; Andres, E.; Hundt, A.; Luzy, A.P. Successful proof of concept of a micronucleus genotoxicity assay on reconstructed epidermis exhibiting intrinsic metabolic activity. *Mutat. Res. Genet. Toxicol. Environ. Mutagenesis* **2018**, *829–830*, 75–86. [CrossRef]
47. OECD. *Test No.231. Guidance Document on the In Vitro Bhas 42 Cell Transformation Assay Series on Testing Assessment*; ENV/JM/MONO(2016)1; OECD Publishing: Paris, France, 2016.
48. Mascolo, M.G.; Perdichizzi, S.; Vaccari, M.; Rotondo, F.; Zanzi, C.; Grilli, S.; Paparella, M.; Jacobs, M.N.; Colacci, A. The transformics assay: First steps for the development of an integrated approach to investigate the malignant cell transformation in vitro. *Carcinogenesis* **2018**, *39*, 955–967. [CrossRef]
49. Pfuhler, S.; van Benthem, J.; Curren, R.; Doak, S.H.; Dusinska, M.; Hayashi, M.; Heflich, R.H.; Kidd, D.; Kirkland, D.; Luan, Y.; et al. Use of in vitro 3D models in genotoxicity testing strategic fit, validation status and way forward. Report of working group of the 7th International workshop on genotoxicity testing (IWGT). *Mutat. Res.* **2020**, *850–851*, 503135. [CrossRef]
50. Kimber, I.; Basketter, D.A.; Gerberick, G.F.; Ryan, C.; Dearman, R.J. Chemical allergy: Translating biology into hazard characterization. *Toxicol. Sci.* **2011**, *120* (Suppl. 1), S238–S268. [CrossRef] [PubMed]
51. Daniel, A.B.; Strickland, J.; Allen, D.; Casati, S.; Zuang, V.; Barroso, J.; Whelan, M.; Régimbald-Krnel, M.J.; Kojima, H.; Nishikawa, A.; et al. International regulatory requirements for skin sensitization testing. *Regul. Toxicol. Pharmacol.* **2018**, *95*, 52–65. [CrossRef]
52. Commission Regulation (EU). 2017/706 of 19 April 2017 amending Annex VII to Regulation (EC) No 1907/2006 of the European Parliament and of the Council on the Registration, Evaluation, Authorisation and Restriction of Chemicals (REACH) as regards

skin sensitisation and repealing Commission Regulation (EU) 2016/1688 (Text with EEA relevance.) C/2017/2369. *Off. J. Eur. Union* **2017**, 8–11. Available online: http://data.europa.eu/eli/reg/2017/706/oj (accessed on 7 June 2021).
53. OECD. *Test No. 406: Skin Sensitisation, OECD Guidelines for the Testing of Chemicals*; Section 4; OECD: Paris, France, 1992. [CrossRef]
54. OECD. *Test No. 429: Skin Sensitisation: Local Lymph Node Assay, OECD Guidelines for the Testing of Chemicals*; Section 4; OECD: Paris, France, 2010. [CrossRef]
55. OECD. *The Adverse Outcome Pathway for Skin Sensitisation Initiated by Covalent Binding to Proteins, OECD Series on Testing and Assessment*; n° 168; OECD: Paris, France, 2014. [CrossRef]
56. OECD. *Test No. 442C: In Chemico Skin Sensitisation: Assays Addressing the Adverse Outcome Pathway Key Event on Covalent Binding to Proteins, OECD Guidelines for the Testing of Chemicals*; Section 4; OECD: Paris, France, 2020. [CrossRef]
57. OECD. *Test No. 442D: In Vitro Skin Sensitisation: ARE-Nrf2 Luciferase Test Method, OECD Guidelines for the Testing of Chemicals*; Section 4; OECD: Paris, France, 2018. [CrossRef]
58. OECD. *Test No. 442E: In Vitro Skin Sensitisation: In Vitro Skin Sensitisation Assays Addressing the Key Event on Activation of Dendritic Cells on the Adverse Outcome Pathway for Skin Sensitisation, OECD Guidelines for the Testing of Chemicals*; Section 4; OECD: Paris, France, 2018. [CrossRef]
59. Kleinstreuer, N.C.; Hoffmann, S.; Alepee, N.; Allen, D.; Ashikaga, T.; Casey, W.; Clouet, E.; Cluzel, M.; Desprez, B.; Gellatly, N.; et al. Non-animal methods to predict skin sensitization (II): An assessment of defined approaches (*). *Crit. Rev. Toxicol.* **2018**, *48*, 359–374. [CrossRef] [PubMed]
60. OECD. Test Guidelines Programme Work Plan. Available online: https://www.oecd.org/chemicalsafety/testing/Test_Guidelines_Workplan_2020.pdf (accessed on 29 April 2021).
61. Roberts, D.W. Is a combination of assays really needed for non-animal prediction of skin sensitization potential? Performance of the GARD (Genomic Allergen Rapid Detection) assay in comparison with OECD guideline assays alone and in combination. *Regul. Toxicol. Pharmacol.* **2018**, *98*, 155–160. [CrossRef] [PubMed]
62. Mehling, A.; Adriaens, E.; Casati, S.; Hubesch, B.; Irizar, A.; Klaric, M.; Letasiova, S.; Manou, I.; Müller, B.P.; Roggen, E.; et al. In vitro RHE skin sensitisation assays: Applicability to challenging substances. *Regul. Toxicol. Pharmacol.* **2019**, *108*, 104473. [CrossRef] [PubMed]
63. Johansson, H.; Lindstedt, M.; Albrekt, A.S.; Borrebaeck, C.A. A genomic biomarker signature can predict skin sensitizers using a cell-based in vitro alternative to animal tests. *BMC Genom.* **2011**, *12*, 399. [CrossRef]
64. Johansson, H.; Gradin, R.; Johansson, A.; Adriaens, E.; Edwards, A.; Zuckerstätter, V.; Jerre, A.; Burleson, F.; Gehrke, H.; Roggen, E.L. Validation of the GARD™ skin assay for assessment of chemical skin sensitizers: Ring trial results of predictive performance and reproducibility. *Toxicol. Sci.* **2019**, *170*, 374–381. [CrossRef]
65. Larne, O.; Mattson, U.; Gradin, R.; Hohansson, H. Extended applicability domain of the GARD platform by solvent-extraction protocols allows for accurate assessment of sensitizing mixtures and UVCBs. In Proceedings of the SOT Annual Meeting 2020, Anaheim, CA, USA, 15–19 March 2020.
66. Johansson, A.; Larne, O.; Pedersen, E.; Berglin, M.; Petersen, H.; Jenvert, R.-M.; Johansson, H. Evaluation of the Applicability of GARDskin to Predict Skin Sensitizers in Leachables from Medical Device Materials. 2021. (Unpublished; Manuscript in Preparation). Available online: https://www.sartorius.com/en/services/validation-service/extractables-leachables-testing?gclid=EAIaIQobChMI64a9jbiP8QIVTNiWCh1ofgGTEAAYASAAEgKNW_D_BwE (accessed on 7 June 2021).
67. Api, A.M.; Basketter, D.A.; Cadby, P.A.; Cano, M.F.; Ellis, G.; Gerberick, G.F.; Griem, P.; McNamee, P.M.; Ryan, C.A.; Safford, R. Dermal sensitization quantitative risk assessment (QRA) for fragrance ingredients. *Regul. Toxicol. Pharmacol.* **2008**, *52*, 3–23. [CrossRef]
68. Basketter, D.; Safford, B. Skin sensitization quantitative risk assessment: A review of underlying assumptions. *Regul. Toxicol. Pharmacol.* **2016**, *74*, 105–116. [CrossRef]
69. Hirota, M.; Fukui, S.; Okamoto, K.; Kurotani, S.; Imai, N.; Fujishiro, M.; Kyotani, D.; Kato, Y.; Kasahara, T.; Fujita, M.; et al. Evaluation of combinations of in vitro sensitization test descriptors for the artificial neural network-based risk assessment model of skin sensitization. *J. Appl. Toxicol.* **2015**, *35*, 1333–1347. [CrossRef] [PubMed]
70. Gradin, R.; Forreryd, A.; Mattson, U.; Jerre, A.; Johansson, H. Quantitative assessment of sensitizing potency using a dose-response adaptation of GARDskin. 2021. (Unpublished; Manuscript in Preparation).
71. Steiling, W. Safety evaluation of cosmetic ingredients regarding their skin sensitization potential. *Cosmetics* **2016**, *3*, 14. [CrossRef]
72. Commission regulation (EU). 2018/605 of 19 April 2018 amending Annex II to Regulation (EC) No 1107/2009 by setting out scientific criteria for the determination of endocrine disrupting properties (Text with EEA relevance). *Off. J. Eur. Union* **2018**, *L101*, 33–36.
73. IPCS Global Assessments of EDCS. Chapter 1: Executive Summary. Available online: https://www.who.int/ipcs/publications/en/ch1.pdf (accessed on 21 May 2021).
74. OECD. *Revised Guidance Document 150 on Standardised Test Guidelines for Evaluating Chemicals for Endocrine Disruption, OECD Series on Testing and Assessment*; n° 150; OECD: Paris, France, 2018. [CrossRef]
75. OECD. *Test No. 455: Performance-Based Test Guideline for Stably Transfected Transactivation In Vitro Assays to Detect Estrogen Receptor Agonists and Antagonists, OECD Guidelines for the Testing of Chemicals*; Section 4; OECD: Paris, France, 2016. [CrossRef]
76. OECD. *Test No. 458: Stably Transfected Human Androgen Receptor Transcriptional Activation Assay for Detection of Androgenic Agonist and Antagonist Activity of Chemicals, OECD Guidelines for the Testing of Chemicals*; Section 4; OECD: Paris, France, 2020. [CrossRef]

77. OECD. *Test No. 456: H295R Steroidogenesis Assay, OECD Guidelines for the Testing of Chemicals*; Section 4; OECD: Paris, France, 2011. [CrossRef]
78. Loughlin, K.R. The clinical applications of five-alpha reductase inhibitors. *Can. J. Urol.* **2021**, *28*, 10584–10588.
79. Rossier, N.M.; Chew, G.; Zhang, K.; Riva, F.; Fent, K. Activity of binary mixtures of drospirenone with progesterone and 17α-ethinylestradiol in vitro and in vivo. *Aquat. Toxicol.* **2016**, *174*, 109–122. [CrossRef] [PubMed]
80. Leusch, F.D.L.; Aneck-Hahn, N.H.; Cavanagh, J.E.; Du Pasquier, D.; Hamers, T.; Hebert, A.; Neale, P.A.; Scheurer, M.; Simmons, S.O.; Schriks, M. Comparison of in vitro and in vivo bioassays to measure thyroid hormone disrupting activity in water extracts. *Chemosphere* **2018**, *191*, 868–875. [CrossRef]
81. OECD. *Test No. 248: Xenopus Eleutheroembryonic Thyroid Assay (XETA), OECD Guidelines for the Testing of Chemicals*; Section 2; OECD: Paris, France, 2019. [CrossRef]
82. Belanger, S.E.; Balon, E.K.; Rawlings, J.M. Saltatory ontogeny of fishes and sensitive early life stages for ecotoxicology tests. *Aquat. Toxicol.* **2010**, *97*, 88–95. [CrossRef]
83. Petersen, K.; Fetter, E.; Kah, O.; Brion, F.; Scholz, S.; Tollefsen, K.E. Transgenic (cyp19a1b-GFP) zebrafish embryos as a tool for assessing combined effects of oestrogenic chemicals. *Aquat. Toxicol.* **2013**, *138–139*, 88–97. [CrossRef]
84. Spirhanzlova, P.; Leleu, M.; Sébillot, A.; Lemkine, G.F.; Iguchi, T.; Demeneix, B.A.; Tindall, A.J. Oestrogen reporter transgenic medaka for non-invasive evaluation of aromatase activity. *Comp. Biochem. Physiol. C Toxicol. Pharmacol.* **2016**, *179*, 64–71. [CrossRef]
85. European Chemical Agency (ECHA); European Food Safety Authority (EFSA); Joint Research Centre (JRC); Andersson, N.; Arena, M.; Auteri, D.; Barmaz, S.; Grignard, E.; Kienzler, A.; Lepper, P.; et al. Guidance for the identification of endocrine disruptors in the context of Regulations (EU) No 528/2012 and (EC) No 1107/2009. *EFSA J.* **2018**, *16*, e05311. [CrossRef]
86. Sébillot, A.; Damdimopoulou, P.; Ogino, Y.; Spirhanzlova, P.; Miyagawa, S.; Du Pasquier, D.; Mouatassim, N.; Iguchi, T.; Lemkine, G.F.; Demeneix, B.A.; et al. Rapid fluorescent detection of (anti)androgens with spiggin-gfp medaka. *Environ. Sci. Technol.* **2014**, *48*, 10919–10928. [CrossRef] [PubMed]
87. OECD. *Test No. 428: Skin Absorption: In Vitro Method, OECD Guidelines for the Testing of Chemicals*; Section 4; OECD: Paris, France, 2004. [CrossRef]
88. OECD. *Guidance Document for the Conduct of Skin Absorption Studies, OECD Series on Testing and Assessment*; n° 28; OECD: Paris, France, 2004. [CrossRef]
89. OECD. *Test No. 156, Guidance Notes for the Estimation of Dermal Absorption Values, OECD Series on Testing and Assessment*; ENV/JM/MONO (2011)36; OECD: Paris, France, 2011.
90. The Scientific Committee on Cosmetic Products and Non-Food Products Intended for Consumers. *Basic Criteria for the In Vitro Assessment of Dermal Absorption of Cosmetic Ingredients*. Updated October 2003; SCCNFP/0750/03. Available online: https://ec.europa.eu/health/archive/ph_risk/committees/sccp/documents/out231_en.pdf (accessed on 7 June 2021).
91. SCCS (Scientific Committee on Consumer Safety). Basic Criteria for the In Vitro Assessment of Dermal Absorption of Cosmetic Ingredients, 22 June 2010. Available online: https://ec.europa.eu/health/scientific_committees/consumer_safety/docs/sccs_s_002.pdf (accessed on 7 June 2021).
92. Mitra, A.; Kim, N.; Spark, D.; Toner, F.; Craig, S.; Roper, C.; Meyer, T.A. Use of an in vitro human skin permeation assay to assess bioequivalence of two topical cream formulations containing butenafine hydrochloride (1%, w/w). *Regul. Toxicol. Pharmacol.* **2016**, *82*, 14–19. [CrossRef] [PubMed]
93. Barbero, A.M.; Frasch, H.F. Effect of Frozen human epidermis storage duration and cryoprotectant on barrier function using two model compounds. *Skin Pharmacol. Physiol.* **2016**, *29*, 31–40. [CrossRef] [PubMed]
94. Wester, R.C.; Christoffel, J.; Hartway, T.; Poblete, N.; Maibach, H.I.; Forsell, J. Human cadaver skin viability for in vitro percutaneous absorption: Storage and detrimental effects of heat-separation and freezing. *Pharm. Res.* **1998**, *15*, 82–84. [CrossRef]
95. Osman-Ponchet, H.; Boulai, A.; Kouidhi, M.; Sevin, K.; Alriquet, M.; Gaborit, A.; Bertino, B.; Comby, P.; Ruty, B. Characterization of ABC transporters in human skin. *Drug Metab. Pers. Ther.* **2014**, *29*, 91–100. [CrossRef] [PubMed]
96. Alriquet, M.; Sevin, K.; Gaborit, A.; Comby, P.; Ruty, B.; Osman-Ponchet, H. Characterization of SLC transporters in human skin. *ADMET DMPK* **2015**, *3*, 34–44. [CrossRef]
97. Fujiwara, R.; Takenaka, S.; Hashimoto, M.; Narawa, T.; Itoh, T. Expression of human solute carrier family transporters in skin: Possible contributor to drug-induced skin disorders. *Sci. Rep.* **2014**, *4*, 5251. [CrossRef] [PubMed]
98. Clerbaux, L.A.; Paini, A.; Lumen, A.; Osman-Ponchet, H.; Worth, A.P.; Fardel, O. Membrane transporter data to support kinetically-informed chemical risk assessment using non-animal methods: Scientific and regulatory perspectives. *Environ. Int.* **2019**, *126*, 659–671. [CrossRef]
99. Rougier, A.; Lotte, C.; Maibach, H.I. In vivo percutaneous penetration of some organic compounds related to anatomic site in humans: Predictive assessment by the stripping method. *J. Pharm. Sci.* **1987**, *76*, 451–454. [CrossRef]
100. Sandby-Møller, J.; Poulsen, T.; Wulf, H.C. Epidermal thickness at different body sites: Relationship to age, gender, pigmentation, blood content, skin type and smoking habits. *Acta Derm. Venereol.* **2003**, *83*, 410–413. [CrossRef]
101. Marrakchi, S.; Maibach, H.I. Biophysical parameters of skin: Map of human face, regional, and age-related differences. *Contact Dermat.* **2007**, *57*, 28–34. [CrossRef] [PubMed]
102. Shriner, D.L.; Maibach, H.I. Regional variation of nonimmunologic contact urticaria: Functional map of the human face. *Skin Pharmacol.* **1996**, *9*, 312–321. [CrossRef]

103. Lampe, M.A.; Burlingame, A.L.; Whitney, J.; Williams, M.L.; Brown, B.E.; Roitman, E.; Elias, P.M. Human stratum corneum lipids: Characterization and regional variations. *J. Lipid Res.* **1983**, *24*, 120–130. [CrossRef]
104. Endringer-Pinto, F.; Bagger, C.; Kunze, G.; Joly-Tonetti, N.; Thénot, J.P.; Osman-Ponchet, H.; Janfelt, C. Visualization of penetration of topical antifungal drug substances through mycosis-infected nails by matrix assisted laser desorption ionization mass spectrometry imaging. *Mycoses* **2020**, *63*, 869–875. [CrossRef] [PubMed]
105. Cosmetic Ingredeint Review. Dermal Penetration, Absorption, and other Considerations for Babies and Infants in Safety Assessments. Available online: https://www.cir-safety.org/sites/default/files/Infskn092014rep-%20final.pdf (accessed on 7 June 2021).
106. Makri, A.; Goveia, M.; Balbus, J.; Parkin, R. Children's susceptibility to chemicals: A review by developmental stage. *J. Toxicol. Environ. Health B Crit. Rev.* **2004**, *7*, 417–435. [CrossRef] [PubMed]
107. Landrigan, P.J.; Garg, A. Chronic effects of toxic environmental exposures on children's health. *J. Toxicol. Clin. Toxicol.* **2002**, *40*, 449–456. [CrossRef] [PubMed]
108. Schwenk, M.; Gundert-Remy, U.; Heinemeyer, G.; Olejniczak, K.; Stahlmann, R.; Kaufmann, W.; Bolt, H.M.; Greim, H.; von Keutz, E.; Gelbke, H.P. Children as a sensitive subgroup and their role in regulatory toxicology. *DGPT Workshop Rep. Arch. Toxicol.* **2003**, *77*, 2–6. [CrossRef]
109. Ficheux, A.S.; Roudot, P.J. *Evaluation Probabiliste de L'exposition de la Population Française Aux Produits Cosmétiques*; LERCCo, UBO: Brest, France, 2017; Available online: https://www.cert-online.biz/sites/cert-online.biz/files/page/fichiers/lercco_table_des_matieres.pdf (accessed on 7 June 2021).
110. Scientific Committee on Consumer Safety (SCCS). *Clarification on Opinion SCCS/1348/10 in the Light of the Danish Clause of Safeguard Banning the Use of Parabens in Cosmetic Products Intended for Children under Three Years of Age*; SCCS/1446/11; European Commission: Brussels, Belgium, 2011.
111. Osman-Ponchet, H.; Alriquet, M.; Kouidhi, M.; Sevin, K.; Gaborit, A. Use of microneedle device to enhance dermal absorption: Study on ex vivo human skin. *J. Dermat. Cosmetol.* **2018**, *2*, 00032. [CrossRef]
112. Osman-Ponchet, H.; Gaborit, A.; Sevin, K.; Bianchi, C.; Linget, J.M.; Wilson, C.E.; Bouvier, G. Preteatment of skin using an abrasive skin preparation pad, a microneedling device or iontophoresis improves absorption of methyl aminolevulinate in ex vivo human skin. *Photodiagn. Photodyn. Ther.* **2017**, *20*, 130–136. [CrossRef]
113. Osman-Ponchet, H.; Gaborit, A.; Kouidhi, M.; Anglars, M.; Marceau-Suissa, J.; Duffy-Roger, O.; Linget, J.M.; Wilson, C.E. Comparison of the effect of skin preparation pads on transepidermal water loss in ex vivo human skin. *Dermatol. Ther. (Heidelb.)* **2017**, *7*, 407–415. [CrossRef]
114. OECD. *Test No. 404: Acute Dermal Irritation/Corrosion, OECD Guidelines for the Testing of Chemicals*; Section 4; OECD: Paris, France, 2015. [CrossRef]
115. OECD. *Test No. 405: Acute Eye Irritation/Corrosion, OECD Guidelines for the Testing of Chemicals*; Section 4; OECD: Paris, France, 2020. [CrossRef]
116. OECD. *Test No. 439: In Vitro Skin Irritation: Reconstructed Human Epidermis Test Method, OECD Guidelines for the Testing of Chemicals*; Section 4; OECD: Paris, France, 2020.
117. OECD. *Test No. 492: Reconstructed Human Cornea-like Epithelium (RhCE) Test Method for Identifying Chemicals not Requiring Classification and Labelling for Eye Irritation or Serious Eye Damage, OECD Guidelines for the Testing of Chemicals*; Section 4; OECD: Paris, France, 2019. [CrossRef]
118. Aberdam, E.; Petit, I.; Sangari, L.; Aberdam, D. Induced pluripotent stem cell-derived limbal epithelial cells (LiPSC) as a cellular alternative for in vitro ocular toxicity testing. *PLoS ONE* **2017**, *12*, e0179913. [CrossRef] [PubMed]
119. Report from the Commission to the European Parliament and the Council. *2019 Report on the Statistics on the Use of Animals for Scientific Purposes in the Member States of the European Union in 2015–2017*; COM/2020/16 final; Brussels, 5.2.2020. Available online: https://op.europa.eu/en/publication-detail/-/publication/04a890d4-47ff-11ea-b81b-01aa75ed71a1 (accessed on 7 June 2021).

Review

Safety of Tattoos and Permanent Make up (PMU) Colorants

Eleni Andreou [1,*], Sophia Hatziantoniou [2], Efstathios Rallis [1] and Vasiliki Kefala [1]

1. Department of Biomedical Sciences, School of Health Sciences and Welfare, University of West Attica, 122 43 Athens, Greece; efrall@otenet.gr (E.R.); valiakef@uniwa.gr (V.K.)
2. Laboratory of Pharmaceutical Technology, Department of Pharmacy, Health Sciences School, University of Patras, 265 04 Patras, Greece; sohatzi@upatras.gr
* Correspondence: elandreou@uniwa.gr

Abstract: The art of tattooing is a popular decorative approach for body decoration and has a corrective value for the face. The tattooing procedure is characterized by placing exogenous pigments into the dermis with a number of needles. The process of creating traditional and cosmetic tattoos is the same. Colorants are deposited in the dermis by piercing the skin with needles of specific shape and thickness, which are moistened with the colorant. Colorants (pigments or dyes) most of the time include impurities which may cause adverse reactions. It is commonly known that tattoo inks remain in the skin for lifetime. It is also a fact that the chemicals that are used in permanent makeup (PMU) colorants may stay in the body for a long time so there is a significant long-term risk for harmful ingredients being placed in the body. Tattoo and PMU colorants contain various substances and their main ingredients and decomposition components may cause health risks and unwanted side effects to skin.

Keywords: tattoos; permanent makeup; PMU; colorants; pigments

Citation: Andreou, E.; Hatziantoniou, S.; Rallis, E.; Kefala, V. Safety of Tattoos and Permanent Make up (PMU) Colorants. *Cosmetics* **2021**, *8*, 47. https://doi.org/10.3390/cosmetics8020047

Academic Editor: Kalliopi Dodou

Received: 15 April 2021
Accepted: 2 June 2021
Published: 7 June 2021

Publisher's Note: MDPI stays neutral with regard to jurisdictional claims in published maps and institutional affiliations.

Copyright: © 2021 by the authors. Licensee MDPI, Basel, Switzerland. This article is an open access article distributed under the terms and conditions of the Creative Commons Attribution (CC BY) license (https://creativecommons.org/licenses/by/4.0/).

1. Introduction

Tattoos have become a very popular form of body and face art in the last two decades. There is evidence that 12 % of Europeans have at least one tattoo on their body especially in the 18–35 age group [1–3]. The most common tattoos are usually made with black ink or have various colors located on almost all areas of the human body. A survey in German-speaking countries showed that in 60% of body tattoos black ink has been used [4]. When we refer to tattoos, this also includes permanent makeup (PMU), which is mostly applied over the face area. PMU is used especially on the periorbital and perioral regions for decorative reasons. PMU colorants are carefully injected with a PMU machine or by "cutting" the skin via a manual PMU procedure into the face, head and body area. These applications are made by using solid or multiple fine needles injecting the color or by placing the colorants via microblades.

Tattoo and PMU artists use tattoo colorant suspensions from different commercial suppliers in this field; chemical substances in these products are not always approved by the European Chemicals Agency (ECHA) or the Food and Drug Administration (FDA) [5]. Until recently there was no limitation on the use of certain chemicals in tattoo inks and in permanent makeup colorants. There is also no analytical method for the detection of metals, polycyclic aromatic hydrocarbons or forbidden colorants [6]. To protect European citizens, many of the hazardous chemicals found in tattoo inks and PMU are restricted in the EU under the Registration, Evaluation, Authorization and Restriction of Chemicals regulation (REACH) will come into effect in several months [3,7]. Analytical methods focus on exactly what causes the health problems, are extremely specific. Although identifying the correct method for detecting metals in tattoo inks is particularly important, the focus should be on the actual amounts of ink injected in the skin [8]. Ink concentration into dermis is an unknown issue that should be investigated because the number of decorative tattoo applications has increased and the only restriction that currently exists covers chemicals that may cause cancer and genetic mutations or is focused on chemicals that are toxic to reproduction as well as skin sensitizers and irritants. If a serious restriction had been necessitated for every kind of tattooing from every country's law, many cases of chronic allergic reactions would have been prevented [9].

The aim of this review is to raise concerns about the necessity of changing regulations about tattoo ink use and ink manufacturers to make tattooing a much safer application with no cutaneous or systemic adverse effects.

2. Methods

The review was based on a thorough search through the literature in the relevant databases (Scopus, PubMed and Google Scholar). The search terms used, were:
1. Tattoo ink, tattoo colorants, tattoo pigments (title or abstract word), PMU pigments (title or abstract word), permanent makeup colorants (title or abstract word).
2. Adverse effects (title or abstract word), hazard (title or abstract word), side effects, complications, equipment, needles safety.
3. Tattooing, permanent makeup, microblading, micropigmentation.

Two different search groups were used. In the first group the terms were related with "or", and among each group with "and". The published material found on these databases and the World Wide Web was reviewed on the topics of safety of tattoos and permanent makeup colorants, side effects of tattoo application, and PMU application and chemical substances that might cause cutaneous adverse effects. These data were analyzed, and the results are discussed below. In Sections 3–10, information is provided about colorants and potential hazards of tattoo and PMU colorants in the human body.

3. Tattoo Inks

Most of tattoo inks are manufactured today in countries with national regulations over the percentages of hazardous ingredients. A short time ago, the manufacturers of inks had no regulations to follow and many of these products caused adverse reactions to the skin. Today the microbiological quality of tattoo ink and PMU products is good, and products manufactured in the EU have high production standards [10]. Ink and PMU should be produced in a sterile environment to be microbiologically stable for several months after opening. This leads to safer tattoo and PMU applications. The composition of these colorants is important because they can cause side effects such as photoallergic, granulomatous, and anaphylactic reactions. Their chemical composition can be a predictive factor of the tattoo reaction after laser treatments as well [11,12].

When finishing the tattoo or PMU application, part of the injected tattoo colorants leave the wounded skin area and an adequate amount of ink stays in the dermis area, which is the target of pigment particles in order to create the permanent result of the tattoo or PMU. The first days after a tattoo or PMU application the skin starts absorbing the colorants which remain for a long time in the injection site. As time passes pigment particles move deeper in the skin creating the permanent character of the tattoo design. This relocation from epidermis, dermis and, sometimes, subcutaneous tissue, changes the color of the tattoo as the years pass. Another unpredictable reaction of the injected tattoo colorant is their migration from the skin through the lymphatic or blood vessel system; this explains why tattoo colorants can be found in lymph nodes near the tattoo area [13].

It is estimated that every tattoo application injects about 1 mg of ink per cm^2. Scientific research in tissue samples shows that polycyclic aromatic hydrocarbons (PAHs), can be found many years after tattooing in regional lymph nodes. Azo or polycyclic compounds from colored tattoos express the same characteristics. These kinds of pigments are designed mainly for industrial use and not to be injected into human skin [14]. Due to adverse reactions, pigment compositions of certain ink colors have changed through the years [15]. For instance, toxic mercuric sulfide that was once used in red tattoos has now been removed from tattoo inks and PMU colorants because of reported skin reactions [16]. Laser removal is also a problem when it involves colorants with iron oxides and titanium dioxide. These chemicals are getting darker under high-powered laser treatments causing disappointment to the persons who wished to remove their tattoo [17]. Dispersive x-ray showed that the elemental composition of commonly used tattoo pigments, contain complex suspensions of many different chemical substances. The main ingredients of these suspensions are tiny solid particles from black, white, or colored pigments. There were also many other unknown chemical substances and a solvent. The small size of these particles is one of the reasons for the tattoo's lifetime existence on the skin [18].

4. PMU Colorants

PMU is made in the same way as a decorative tattoo application. The skin's outer layer is penetrated with a needle from a PMU machine or a manual PMU pen (microblading) with various numbers of needles stuck together. The colorant is injected or placed into the area beneath, with a needle angle usually at 80°–90°, to make a new brow or correct the brow shape by creating brow hairs or by shading the area. The same procedure is followed for eyebrow reconstruction and modification, in the eyelid area for permanent eyeliner makeup and over the lips area in order to create a better shape and a more vivid color to the lips. The epidermis is regenerated continuously to make the PMU last and the color is injected into the dermis which is about 1.0–2.5 mm deep [19]. The application of permanent makeup is the same as the tattoo procedure and is used to produce designs that look like the aesthetic application of makeup. The quality of the results is determined by the longevity of color and the depth of colorant penetration. When the colorants are placed closed to the epidermis, they "disappear" easily after several months from application. The pigments used for PMU differ according to the procedure used. The PMU machine gives better results with inorganic colors while the manual PMU has a more "natural" result with organic colors. Every tattoo or PMU artist has to be informed about the different types of pigments and their ingredients in order to choose the right pigment color based on the technique used and the skin tone [20].

Medical tattooing is also a PMU procedure. It is used to camouflage scars or to mimic hair or nipple/areola regions after breast CA surgery, and to create micropigmentation for hair loss. PMU machines are used for medical applications because they have better results. Nowadays the application of scalp micropigmentation (SMP) in order to avoid hair transplantation has gained popularity. This technique is based on the application of PMU colors to the skin above the head area in both sexes. The micropigmentation procedure includes the use of a PMU machine, insertion of the needle at 90° to the skin by shading the hair area with a dot-to-dot technique. This kind of application is made to mimic the hair follicles. When the procedure is complete, the head hair looks more plentiful and if the color is similar to the hair color, there is little difference from the natural hair. Similarly, scalp alopecia (total or partial) or scalp scars can be camouflaged with a stippling pattern of pigments that mimic the hair follicles. Due to the specificity of SPM and PMU, colorants should not fade easily as the head and face area are exposed daily to sunlight [21,22].

PMU colors have a variety of pigments that are safe, hypoallergenic and can be used either with the PMU machine or manual PMU pen (microblading) technique. They have a wide range of colors and shades that can be mixed with some restrictions in order to modify to suit the skin color [23,24]. The term pigment is used to define both the fine powder that gives color to cosmetic products such as regular makeup, as well as the solution made by adding these powders to a binder used in permanent makeup. PMU pigments are made by colored liquid concoctions. These colors stay in the skin for a period of time because of the particle size (bigger than that used in tattoo ink), which the body eventually breaks down and absorbs. This PMU application lasts from several months up to a few years so it can be applied again. PMU colorants are one of the differences between permanent makeup pigments and tattoo ink because ink has tiny particles that cannot be broken down; the ink is placed deeper into the skin during the tattoo procedure with different kind of machines.

The tattoo will last a lifetime no matter how much the color changes over the years due to sun exposure and various other reasons [25,26].

5. Colorants

Colorants belong to the same category as pigments or dyes with molecules that have the same chemical structure. On the other hand, dyes have pigments, which are practically insoluble in the medium in which they are incorporated. Although tattooing and PMU are injectable procedures, colorant suspensions are not pharmaceutical substances and do not have the injectable products standards. They contain over 100 different chemical compounds to different extents [27]. They are a mix of several chemicals and may contain hazardous substances that cause skin allergies and other serious health impacts, such as genetic mutations and cancer [13,28]. As the tattoo and PMU procedure are made by the injection in skin of such colorants, they may create health risks on every tattooed individual. Ink pigments have been found in lymph nodes and the liver after migration from the skin area [29]. The choice of pigment type depends on the PMU technique used, the artist's preference, the area of the treatment and the skin type of the client. Most of the tattoo and PMU colorants are made up of both organic and inorganic pigments. The chemical synthesis of pigments has many tiny insoluble particles, which have diameters from a few tenths of nanometers (nanoparticles) up to a few micrometers. Other kinds of finishing processes give the surfaces of the pigment particles the ability for different applications [30,31].

6. Inorganic Pigments

The main characteristic of inorganic pigments is the addition of iron oxide elements. As they are synthetically produced from metals, they have an inorganic character (clay, ultramarines, titanium oxide, manganese violet). They are used in tattoo and PMU applications with the use of a machine but not with the manual PMU pen (microblading). The purpose of adding iron oxides to tattoo and PMU colorants is to provide solid color and opacity and widen the shade range. Titanium dioxide prevails in lighter shades, while iron oxides prevail in darker shades. Titanium dioxide is used as a brightening agent in tattoo pigments, sunscreens and generally in paints. It has a white color and one of its most important characteristics is the absorption of UV light in the range between 280 and 400 nm [7,32].

Today there are a variety of inorganic pigments based on iron oxides in many colors of tattoo ink and PMU colorants. These colors are yellow, red and black, which are based on heavy metals such as mercury sulfide (red), cadmium sulfide (yellow), chromium oxide (green), or cobalt spinel (blue). They are unaffected by light, non-toxic and insoluble, which is important for the prevention of color migration. Inorganic pigments are the least likely to cause an allergic reaction and are a widely used group of pigments in permanent makeup application. This happens because they give a more stable result when they are used for shading techniques because of the tiny pigment particle distribution [33].

7. Organic Pigments

Tattoo colorants contain more than 80% of industrial organic pigments [7,34]. Organic pigments can be produced in a big variety of color shades ranging from green, blue, red and violet to yellow. These kinds of pigments absorb the light resulting in high color strength and have a vivid color in the skin, which lasts for a long time [35]. This characteristic makes these colorants very important for tattoo application. Polycyclic or azo pigments, as they are known, are used in tattoo colorants and are classified by their chemical constitution. Their subdivision is: mono-azo, dis-azo, b-naphtol, and naphthol AS. There are also pigments with metal complexes, which have cobalt, copper and nickel [5,36].

Heterocyclic and aromatic compounds are the characteristics of polycyclic pigments; quinacridone pigments (red, bluish red, violet) and the phthalocyanines (green, blue) are representative examples [37]. The production of purely organic pigments is very small and is performed by a few companies in the market. The reason for this is because this kind of pigment has a complex chemical synthesis, and the colorants contain many by-products along with titanium dioxide [38]. In general, carbon is the basis of organic chemistry, so these are basically carbon derivatives. In the past, they were obtained from plant and animal organisms but that was not a safe option. This is because vegetable dyes can cause allergic reactions in many ways. Today's color production combines carbon with other substances such as oxygen and hydrogen [39].

Changing the ratios modifies color density. Hydroxide aluminum is also a substance used in tattoo and PMU colors. The main characteristic of hydroxide aluminum is that it is not a soluble substance and this leads to color retention. The pigment becomes heavier and that is the reason why it can set into the skin in a better way. The hypoallergenic formulation of the organic pigments which are manufactured today, is given by alumina hydroxide, which creates a protective membrane over the pigment molecules to prevent a direct reaction with the skin tissue. These pigments are called lake pigments [40]. Organic pigments are affected by sunlight exposure and fade easier, a characteristic, which is the reason why they are more often used in PMU applications. PMU is characterized as a semi-permanent application in contrast with the permanent tattoo application. Elemental carbon molecules are the smallest of all ingredients used in PMU; this characteristic gives a pitch-black, opaque color. Although they can be used in PMU applications, they have a high migration risk because of the small particle size [41,42]. In the last few years water-based colorants have appeared. They have no iron oxides and contain around 45% water and are reported as purely botanic [43]. They are used for PMU applications and have a good result over the face area especially on oily skin. They are characterized as "vegan" pigments and they have gained ground in the PMU colorants market.

In Figure 1 we can see the difference between tattoo inks and PMU colors, which can be seen after diluting 1 mL of the colorant in water. The tiny particles of tattoo ink are spread over the glass while the PMU colorant stays at the bottom of the glass and does not mix with the water.

Figure 1. (a) Tattoo ink diluted in water, (b) PMU color diluted in water. Image courtesy of EleniAndreou.

8. Potential Hazards of Tattoo Colorants

Although tattoo and PMU colorants are being injected in the human body to a depth of 1 mm to 3 mm (Figure 2), they have no pharmaceutical guidelines referring to subcutaneous use. They cannot be categorized as cosmetic products or medicines. The exact list of ingredients, if they are referred to at all, depends on the legislation of each country about the manufacturers or importers of this kind of product, so there are significant gaps in our knowledge about their ingredients. Tattoo colorants containing hazardous chemicals have been found on the European market. In samples taken from the tattoo and PMU colorant market, microbiological contamination, heavy metals, polycyclic aromatic hydrocarbons (PAH), primary aromatic amines (PAA), and preservatives were found [7]. As the application of tattoo and PMU is administered invasively and creates skin injury, the healing process should last a few days. A survey of 3411 tattooed participants revealed that 8% of the participants still had health problems 4 weeks after tattooing, and 6% had persistent skin problems in the tattooed area [4,5]. There is a correlation of these problems with the tattoo ink, the colors and their ingredients. This was also confirmed by comparing the data of the survey with medical case reports. The results showed that colored tattoo pigments have cutaneous effects over the skin area [44].

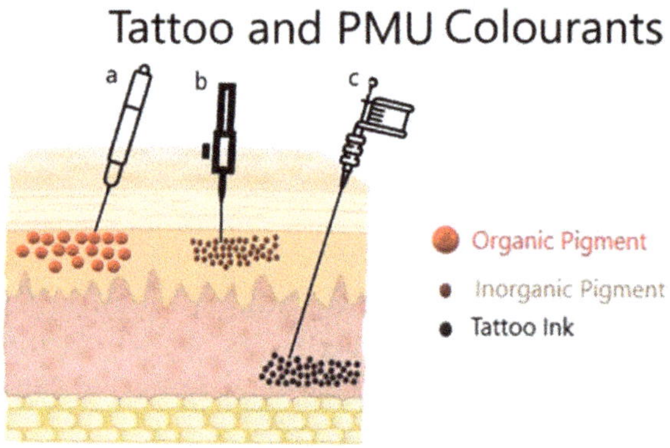

Figure 2. The depth of tattoo and PMU colorants according to organic and inorganic ink and different machine use. Inorganic pigments contain bigger particles of colorants and are placed deeper into the skin. (**a**) Manual PMU pen (microblading), (**b**) PMU machine, (**c**) tattoo machine. Image courtesy of Eleni Andreou.

As there are large amounts of PAH in black colorants, hazardous substances are being injected into the skin. This observation is verified by ink found in human organs even in placenta [45]. Generally, there is a lack of scientific investigations or epidemiologic data about the systemic effects of tattoo colorants and their decomposition products. The tattoo trend has led to millions of people having many, often large, tattoos over their body. The majority of these have a size of 600 cm^2 or more. Such a tattoo can include about 1500 mg of azo pigments. These pigments are injected into the human body and there is the possibility of skin or internal organ health problems [4,46].

9. Ink Market

Tattoo and PMU ink markets are spread all over the world. It is estimated that 80% of tattoo inks are manufactured outside of Europe. Asian and American products used by tattoo artists have a dominant presence in the market. The pricing policy of these products vary in every country, with the Asian market being the most competitive. A percentage of 70% of permanent makeup inks are manufactured in Europe, but also many products are imported from America and Asia. In Europe, tattoo inks and PMU colorants are manufactured by about 30 companies located in Spain, Netherlands, Germany, Italy, England and France [47]. In 2008, the European Council Resolution ResAP (2008)1 was created regarding the requirements and criteria for the safety of tattoos and permanent makeup guides for the manufacturers of tattoo inks. Currently, Netherlands, Norway, France, Germany, Spain, Switzerland and Sweden have adopted national regulations on tattoo ink manufacturing and Austria, Italy, Denmark and Slovenia are using the resolution to control tattoo inks [48]. Today ink manufacturers should follow the rules of the Classification, Labelling and Packaging (CLP) Regulation in terms of labelling products that contain classified substances in excess of their classification limits and REACH in terms of registration requirements and information provision. It is still unknown what happens when ink enters the body and this need led to a prohibited ingredients list creation [49]. It is commonly accepted that forbidden ingredients for cosmetic use products cannot be used in tattoo inks and PMU colorants. Ingredients with no cosmetic regulation can be used. A prohibited list with dangerous colorants has been created in order to protect consumers [50]. This year there was an improvement according to a recent article in *The Brussels Times*. The European Chemicals Agency is working on a union-wide bill that will tackle the use of "CMR substances: carcinogenic or causing cancer, mutagenic or affecting cell development and reprotoxic, which interfere with fertility and the reproductive system" in tattooing and PMU [51]. In several months, a more condensed list of hazardous chemicals found in tattoo inks and permanent makeup under restriction by the REACH Regulation, is going to be announced in order to protect European citizens. The target is not banning tattooing procedures, but to make the color use safer [3].

10. Conclusions

The increasing popularity of tattoos and PMU applications expose human skin to colorants and chemical substances that can lead, in many cases, to skin and health problems. The complexity of tattoo ink compounds includes organic dyes, metals, and solvents which might have a hazardous effect on the human body. The unclear identification of tattoo inks as cosmetic products or medicines although they are injected into the skin without being authorized as sterile and injectable, is a serious issue. Recent studies show that 28% of tattooed individuals have more than four tattoos on their body, including PMU applications. The safety of tattoo inks and PMU colorants has obviously increased in Europe in the last few years. From the creation of the European Council Resolution ResAP (2008)1, which resulted in the improved quality control of pigment raw materials, significant changes have been achieved. Further scientific investigation over the tattoo issue needs to be performed. The necessity to explore the unknown long-term side effects of various inks and colorants into the skin is huge, and every country should create a strict regulation about tattoo and PMU applications. It seems obvious that important steps should be obtained over the regulations of the colorant production, which represent the basic elements of tattooing. Furthermore all tattoo artists should be properly trained to identify harmful ingredients written on the product packaging. Each country should be responsible for every tattoo and permanent makeup studio and provide lists with forbidden substances in colorants and perform random checks on ink imports.

Author Contributions: All authors (E.A., S.H., E.R., V.K.) conceived and designed the review; E.A. wrote the paper; E.R. and V.K. supervised the paper. All authors have read and agreed to the published version of the manuscript.

Funding: This review received no external funding.

Institutional Review Board Statement: Not applicable.

Conflicts of Interest: The authors declare no conflict of interest.

References

1. Giulbudagian, M.; Schreiver, I.; Singh, A.V.; Laux, P.; Luch, A. Safety of tattoos and permanent make-up: A regulatory view. *Arch. Toxicol.* **2020**, *94*, 357–369. [CrossRef]
2. Nho, S.W.; Kim, M.; Kweon, O.; Kim, S.J.; Moon, M.S.; Periz, G.; Huang, M.-C.J.; Dewan, K.; Sadrieh, N.K.; Cerniglia, N.K. Microbial contamination of tattoo and permanent makeup inks marketed in the US: A follow-up study. *Lett. Appl. Microbiol.* **2020**, *71*, 351–358. [CrossRef]
3. European Chemicals Agency. Available online: https://echa.europa.eu/hot-topics/tattoo-inks. (accessed on 10 April 2021).
4. Klugl, I.; Hiller, K.A.; Landthaler, M.; Baumler, W. Incidence of health problems associated with tattooed skin: A nation-wide survey in German-Speaking countries. *Dermatology* **2010**, *221*, 43–50. [CrossRef] [PubMed]
5. Bäumler, W. Chemical hazard of tattoo colorants. *Presse Med.* **2020**, *49*, 104046. [CrossRef]
6. Kluger, N. Epidemiology of tattoos in industrialized countries. *Curr. Probl. Dermatol.* **2015**, *48*, 6–20.
7. Piccinini, P.; Pakalin, S.; Contor, L.; Bianchi, I.; Senaldi, C. *Safety of Tattoos and Permanent Make-Up: Final Report*; JRC Science for policy report EUR 27947; European Commission: Brussels, Belgium, 2016.
8. Kluger, N. Cutaneous complications related to permanent decorative tattooing. *Expert Rev. Clin. Immunol.* **2010**, *6*, 363–371. [CrossRef] [PubMed]
9. Kluger, N. Cutaneous infections related to permanent tattooing. *Med. Mal. Infect.* **2011**, *41*, 115–122. [CrossRef] [PubMed]
10. Petersen, H.; Lewe, D. Chemical purity and toxicity of pigments used in tattoo inks. *Curr. Prob. Dermatol.* **2015**, *48*, 136–141.
11. Ortiz, A.E.; Avram, M.M. Redistribution of ink after laser tattoo removal. *Dermatol. Surg.* **2012**, *38*, 1730–1731. [CrossRef] [PubMed]

12. Dirks, M. Making innovative tattoo ink products with improved safety: Possible and impossible ingredients in practical usage. *Curr. Prob. Dermatol.* **2015**, *48*, 118–127.
13. Kluger, N.; Koljonen, V. Tattoos, inks, and cancer. *Lancet Oncol.* **2012**, *13*, 161–168. [CrossRef]
14. Laux, P.; Tralau, T.; Tentschert, J.; Dahouk, S.A.; Bäumler, W.; Bernstein, E.; Bocca, B.; Alimonti, A.; Colebrook, H.; de Cuyper, C.; et al. A medical-toxicological view of tattooing. *Lancet* **2016**, *387*, 395–402. [CrossRef]
15. Kluger, N. Contraindications for tattooing. *Curr. Probl. Dermatol.* **2015**, *48*, 76–87. [PubMed]
16. Gaudron, S.; Ferrier-LeBouëdec, M.C.; Franck, F.; D'Incan, M. Azo pigments and quinacridones induce delayed hypersensitivity in red tattoos. *Contact Dermat.* **2014**, *72*, 97–105. [CrossRef]
17. Hutton Carlsen, K.; Serup, J. Patients with tattoo reactions have reduced quality of life and suffer from itch: Dermatology Life Quality Index and Itch Severity Score measurements. *Ski. Res. Technol.* **2015**, *21*, 101–107. [CrossRef]
18. Engel, E.; Vasold, R.; Santarelli, F.; Maisch, T.; Gopee, N.; Howard, P.; Landthaler, M.; Bäumler, W. Tattooing of skin results in transportation and light-induced decomposition of tattoo pigments—A first quantification in vivo using a mouse model. *Exp. Dermatol.* **2010**, *19*, 54–60. [CrossRef]
19. De Cuyper, C. Complications of Cosmetic Tattoos. *Curr. Probl. Dermatol.* **2015**, *48*, 61–70.
20. De Cuyper, C. Cosmetic and Medical Applications of Tattooing. In *Dermatologic Complications with Body Art*; Cuyper, C.D., Pérez-Cotapos, S.M.L., Eds.; Springer: Berlin/Heidelberg, Germany, 2009.
21. Seyhan, T.; Kapi, E. Scalp Micropigmentation Procedure: A Useful Procedure for Hair Restoration. *J. Craniofacial Surg.* **2021**, *32*, 1049–1053. [CrossRef]
22. Saed, S.; Ibrahim, O.; Bergfeld, W. Hair camouflage: A comprehensive review. *Int. J. Women's Dermatol.* **2017**, *3*, 75–80. [CrossRef]
23. Bäumler, W. Absorption, distribution, metabolism and excretion of tattoo colorants and ingredients in mouse and man: The known and the unknown. *Curr. Prob. Dermatol.* **2015**, *48*, 176–184.
24. Wenzel, S.M.; Welzel, J.; Hafner, C.; Landthaler, M.; Bäumler, W. Permanent make-up colorants may cause severe skin reactions. *Contact Dermat.* **2010**, *63*, 223–227. [CrossRef]
25. Andreou, E.; Hatziantoniou, S.; Rallis, E.; Kefala, V. Legislation and Side Effects induced by Permanent Make Up Colors. *Ep. Klin. Farmakol. Kai Farmakokinet.* **2020**, *38*, 195–201.
26. Biskanaki, F.; Kefala, V. New strategies in Cosmetic Tattoo (Permanent Makeup) and Tattoo removal. *Rev. Clin. Pharmacol. Pharmacokinetics. Int. Ed.* **2018**, *32*, 17–21.
27. Guerra, E.; Llompart, M.; Garcia-Jares, C. Analysis of Dyes in Cosmetics: Challenges and Recent Developments. *Cosmetics* **2018**, *5*, 47. [CrossRef]
28. Schreiver, I.; Hutzler, C.; Andree, S.; Laux, P.; Luch, A. Identification and hazard prediction of tattoo pigments by means of pyrolysis-gas chromatography/mass spectrometry. *Arch. Toxicol.* **2016**, *90*, 1639–1650. [CrossRef]
29. Andreou, E.; Kefala, V.; Rallis, E. Why do cosmetic tattoos change color. An update. *Rev. Clin. Pharmacol. Pharmacokinet. Int. Ed.* **2018**, *32*, 115–123.
30. De Cuyper, C.; Lodewick, E.; Schreiver, I.; Hesse, B.; Seim, C.; Castillo-Michel, H.; Laux, P.; Luch, A. Are metals involved in tattoo-related hypersensitivity reactions? A case report. *Contact Dermat.* **2017**, *77*, 397–405. [CrossRef]
31. Zhang, M.; Jin, J.; Chang, Y.N.; Chang, X.; Xing, G. Toxicological properties of nanomaterials. *J. Nanosci. Nanotechnol.* **2014**, *14*, 717–729. [CrossRef]
32. IARC. Titanium dioxide. *IARC Monogr.* **2010**, *93*, 193–275.
33. Engel, E.; Santarelli, F.; Vasold, R. Modern tattoos cause high concentrations of hazardous pigments in skin. *Contact Dermat.* **2008**, *58*, 228–233. [CrossRef]
34. De Cuyper, C. Tattoo allergy. Can we identify the allergen? *Presse Med.* **2020**, *49*, 104047. [CrossRef]
35. Prior, G. Tattoo inks: Legislation, pigments, metals and chemical analysis. *Curr. Probl. Dermatol.* **2015**, *48*, 152–157.
36. Hauri, U. *Inks for Tattoos and PMU (Permanent Make-Up)/Organic Pigments, Preservatives and Impurities Such as Primary Aromatic Amines and Nitrosamines*; State Laboratory of the Canton of Basel-Stadt: Basel, Switzerland, 2011; pp. 1–10.
37. Bäumler, W.; Eibler, E.T.; Hohenleutner, U.; Sens, B.; Sauer, J.; Landthaler, M. Q-switch laser and tattoo pigments: First results of the chemical and photophysical analysis of 41 compounds. *Lasers Surg. Med.* **2000**, *26*, 13–21. [CrossRef]
38. Hogsberg, T.; Loeschner, K.; Lof, D.; Serup, J. Tattoo inks in general usage contain nanoparticles. *Br. J. Dermatol.* **2011**, *165*, 1210–1218. [CrossRef] [PubMed]
39. Bonadonna, L. Survey of studies on microbial contamination of marketed tattoo inks. *Curr. Probl. Dermatol.* **2015**, *48*, 190–195. [PubMed]
40. Nho, S.W.; Kim, S.J.; Kweon, O.; Howard, P.C.; Moon, M.S.; Sadrieh, N.K.; Cerniglia, C.E. Microbiological survey of commercial tattoo and permanent makeup inks available in the United States. *J. Appl. Microbiol.* **2018**, *124*, 1294–1302. [CrossRef]
41. Lehner, K.; Santarelli, F.; Vasold, R.; König, B.; Landthaler, M.; Bäumler, W. Black tattoo inks are a source of problematic substances such as dibutyl phthalate. *Contact Dermat.* **2011**, *65*, 231–238. [CrossRef] [PubMed]
42. Schreiver, I.; Hesse, B.; Seim, C.; Castillo-Michel, H.; Villanova, J.; Laux, P.; Dreiack, N.; Penning, R.; Tucoulou, R.; Cotte, M.; et al. Synchrotron-based ν-XRF mapping and μ-FTIR microscopy enable to look into the fate and effects of tattoo pigments in human skin. *Sci. Rep.* **2017**, *7*, 11395. [CrossRef]
43. Liszewski, W.; Warshaw, E.M. Pigments in American tattoo inks and their propensity to elicit allergic contact dermatitis. *J. Am. Acad. Dermatol.* **2019**, *81*, 379–385. [CrossRef]

44. Wenzel, S.M.; Rittmann, I.; Landthaler, M.; Bäumler, W. Adverse reactions after tattooing: Review of the literature and comparison to results of a survey. *Dermatology* **2013**, *226*, 138–147. [CrossRef] [PubMed]
45. Ren, A.; Qiu, X.; Jin, L.; Ma, J.; Li, Z.; Zhang, L.; Zhu, H.; Finnell, R.H.; Zhu, T. Association of selected persistent organic pollutants in the placenta with the risk of neural tube defects. *Proc. Nat. Acad. Sci. USA* **2011**, *108*, 12770–12775. [CrossRef] [PubMed]
46. Heywood, W.; Patrick, K.; Smith, A.M.A.; Simpson, J.M.; Pitts, M.K.; Richters, J.; Shelley, J.M. Who Gets Tattoos? Demographic and Behavioral Correlates of Ever Being Tattooed in a Representative Sample of Men and Women. *Ann. Epidemiol.* **2012**, *22*, 51–56. [CrossRef]
47. Michel, R. Manufacturing of Tattoo Ink Products Today and in Future: Europe. *Curr. Probl. Dermatol.* **2015**, *48*, 103–111. [PubMed]
48. Council of Europe. *Committee of Ministers: Resolution ResAP (2008)1 on Tattoos and Permanent Make-Up*; Council of Europe: Strasbourg, France, 2008.
49. Eurofins-Chem-MAP. Available online: https://www.chem-map.com/chemical_news/seven-eu-member-states-introduce-national-legislation-on-tattoo-ink-chemicals-and-permanent-make-up/ (accessed on 10 April 2021).
50. Cosmetic Products Regulation, Annex II—Prohibited Substances. Available online: https://www.echa.europa.eu/web/guest/cosmetics-prohibited-substances (accessed on 15 April 2021).
51. Brussels Times: EU to Tackle Hazardous Chemicals in Permanent Makeup and Tattoo Ink. Available online: https://www.brusselstimes.com/news/art-culture/146340/eu-to-tackle-hazardous-chemicals-in-permanent-makeup-and-tattoo-ink/ (accessed on 10 April 2021).

Technical Note

Designing a Suitable Stability Protocol in the Face of a Changing Retail Landscape

Laura Kirkbride *, Laura Humphries, Paulina Kozielska and Hannah Curtis

Orean Personal Care Ltd., Cleckheaton BD19 4TT, West Yorkshire, UK; laura.humphries@orean.co.uk (L.H.); paulina.kozielska@orean.co.uk (P.K.); hannah.curtis@orean.co.uk (H.C.)
* Correspondence: laura.kirkbride@orean.co.uk

Citation: Kirkbride, L.; Humphries, L.; Kozielska, P.; Curtis, H. Designing a Suitable Stability Protocol in the Face of a Changing Retail Landscape. *Cosmetics* **2021**, *8*, 64. https://doi.org/10.3390/cosmetics8030064

Academic Editor: Kalliopi Dodou

Received: 25 May 2021
Accepted: 30 June 2021
Published: 2 July 2021

Publisher's Note: MDPI stays neutral with regard to jurisdictional claims in published maps and institutional affiliations.

Copyright: © 2021 by the authors. Licensee MDPI, Basel, Switzerland. This article is an open access article distributed under the terms and conditions of the Creative Commons Attribution (CC BY) license (https://creativecommons.org/licenses/by/4.0/).

Abstract: Many recommended stability practices have been unchanged for decades and yet the retail landscape has considerably evolved during that time. First, as a result of the rise of social media and second in the wake of the COVID-19 global pandemic. This article reviews the published guidelines available to the cosmetic scientist when developing a suitable stability protocol and considers them in the context of a changing retail landscape. It sets the context with a background to stability testing and a summary of the relevant regulations across different territories. It outlines the current recommended guidelines for stability testing as stated in publications, including the International Federation of the Societies of Cosmetic Chemists (IFSCC) monograph and Cosmetics Europe. Modern advances in stability testing are also considered including early stability prediction techniques. The article concludes that accelerated stability testing is not a precise science, rather a prediction of shelf life. Scientists must consider the various modes of transport, sizes of shipments and regulation in the country of destination as well as the new and emerging ways of consumer production interaction when developing a suitable stability protocol for their formulation.

Keywords: stability testing; stability protocol; accelerated ageing; shelf life; minimally disruptive formulas; direct to consumer; retail model

1. Introduction

Stability testing or product safety testing encompasses many different aspects of a product formulation. There needs to be proof that the preservative system is efficient, that the product is physically stable, and that the product does not negatively interact with the packaging. Preservative systems are tested through challenge testing, where samples of a product are inoculated with different types of bacteria [1]. The physical product stability is tested by placing samples in inert glass jars and subjecting them to different environment conditions, this type of testing is the main focus of this paper. Product packaging is tested by filling the agreed upon final packaging with the product and again subjecting it to different environment conditions, further testing is also done to assess the packaging functionality with the type of product.

There are many guidelines as to protocols that can be followed when designing a suitable stability testing regime from the established to the more recent thinking. This paper will examine those recommended guidelines in the face of a changing retail landscape. It will explore the International Federation of the Societies of Cosmetic Chemists (IFSCC) monograph and recommendations from Cosmetics Europe, through to more recent work such as that of the UK-based Centre for Process Information (CPI). There have been significant changes in the retail landscape over the decades as the direct-to-consumer (DTC) model has gained prominence. In addition to the DTC model there has been a shift in so called green consumer behaviour, with consumers and producers being more open to the usage of natural cosmetics [2]. This can give rise to additional challenges in stabilising formulations as considered by Singh et al. with their work examining the carrot seed oil-based emulsions [3]. For the purpose of this paper, however, all cosmetic preparations will be considered equally. The paper seeks to examine the protocols available and consider their suitability in the face of the shift in retail behaviour.

2. A Background to Stability Testing

Stability is defined in Cambridge dictionary as a situation in which something is not likely to move or change [4]. Transposing this to a context of a cosmetic product, this can be defined as remaining within a set specification. Specification might include various characteristics like appearance, pH, viscosity or efficacy. They need to be measurable and applicable to a product type. An important specification attribute is microbial stability. Ensuring there is no growth of bacteria or other microorganisms in the product throughout its shelf life is one of the qualities ensuring consumers' safety. Stability assessment preserves the reputation of the brands by making sure products are aesthetically acceptable for consumers [5,6]. Testing of stability helps to establish shelf life of a product during which product continues to be safe and fit for use.

As defined by European Cosmetic Regulation (EC) No 1223/2009, Part A of a Cosmetic Product Safety Report shall contain physical/chemical characteristics and stability of a cosmetic product. Part B of the report which is a safety assessment requires to consider impacts of stability on the safety of cosmetic product. However, the regulation does not specify requirements of how a stability test should be performed. A stability test is essential to evaluate a product's shelf life. There are stipulations in Article 19 of the Regulation about clearly labelling a product with the date of minimum durability if this is less than 30 months. For products with a minimum durability date over 30 months an indication of the period of time after opening (PAO) shall be made instead [7]. No specific instruction on how to calculate PAO is provided in the Regulation. However, microbiological stability as well as packaging of a product should be considered during PAO determination process.

Following the exit of the United Kingdom from European Union in 2020, cosmetic products in the UK are regulated by Schedule 34 of The Product Safety and Metrology, etc. (Amendment, etc.) (EU Exit) Regulations 2019, also called the UK Cosmetics Regulation. The stipulations regarding stability testing are the same as of (EC) No 1223/2009 [8].

In the United States of America cosmetic products are regulated by Food and Drug Administration (FDA) with the Federal Food, Drug and Cosmetic Act (FD&C Act). The FD&C Act prohibits the distribution of cosmetics which are adulterated or misbranded. FD&C Act does not state any requirements regarding product stability or shelf life [9]. However, as the product must be safe it is a responsibility of the manufacturer to verify product's shelf life [10]. Products which fall under the category of an Over-the-Counter Drug in US, like sunscreen or anti-acne treatments, must conform to Current Good Manufacturing Practice for Finished Pharmaceuticals that provides some specifics regarding testing [11]. Further details on storage conditions, testing frequency and more are provided in a supplementary guidance for the industry in Q1A(R2) Stability Testing of New Drug Substances and Products [12]. Comparably to the US, Canada does not set rules or guidelines for stability testing of cosmetic products but defines set of guidelines for drug products in Guidance for Industry Stability Testing of New Drug Substances and Products [13].

Association of Southeast Asian Nations (ASEAN) market, primarily following EU Cosmetic Regulation, in the Guidelines for the safety assessment of a cosmetic product specifies stability needs to be considered [14]. Stability needs to be provided in Part III: Quality Data of Finished Product as part of Product Information File [15]. However, no exact details on test protocol are provided.

3. Current Recommended Guidelines for Stability Testing

Due to the extremely wide variety of products produced in the personal care industry there is no single stability testing procedure that is required for manufacturers to follow when producing a new cosmetic product. Alternatively, there have been a number of recommended guidelines published by global cosmetic associations such as the IFSCC and Cosmetics Europe, these publications suggest protocols to follow when carrying out stability testing. In 2018 the British Standards Institution published ISO/TR 18811:2018 Cosmetics–Guidelines on the stability testing of cosmetics [16]. The document does not aim to specify how a stability test should be performed but does review the information provided in previously published documents on stability testing such as the documents published by the IFSCC and Cosmetic Europe. The ISO is a good starting resource to help manufacturers select the correct protocols for designing a stability test.

The IFSCC is a worldwide federation whose purpose is to promote international cooperation within the personal care industry [17]. The IFSCC published a monograph in 1992 titled "The Fundamentals of Stability Testing" [18] which covers the types of tests and varied conditions that they recommend being used when performing a stability test.

Cosmetics Europe are a trade association specifically for personal care manufacturers within Europe, they are a membership-based association and provide expert knowledge on European legislations and developments within the industry [19]. In 2004 Cosmetics Europe published a report titled "Guidelines on Stability Testing of Cosmetic Products" [20] which sets out guidelines to predict and guarantee the stability of cosmetic products in the market. Both publications by the IFSCC and Cosmetics Europe set out similar guidelines as there are certain tests that have been identified as the best way to test stability, however it is usually down to the manufacturer to design the specifics of the test, hence the need for guidelines.

There are multiple different reasons why a product would need to be stability tested, the IFSCC recommends that the purpose of the test should be identified prior to starting testing so that a sufficient test procedure can be followed. Some of the reasons why a product would require stability testing include, assessment of a new product development (NPD) formulation, assessment of an NPD formulation with its packaging, if a method or formulation has been modified from the original, or if the product container changes [20].

The main test the IFSCC cover is the standard stability test, this is the fundamental test that cosmetic products undergo when performing stability testing. Samples are placed in different temperature environments and the product reactions to these conditions are observed over a set amount of time. Warmer temperature conditions will accelerate any reactions that may occur under normal shelf-life conditions, at a molecular level approximately a 10 °C temperature increase will double the rate of reaction. However, this rule does not accurately apply to more complex systems such as cosmetics, but an increased reaction rate of any amount is beneficial in personal care as it is a fast-moving consumer goods (FMCG) industry. The shelf life of a product can last up to 2–3 years, but the time frame of the development of a product from brief to launch is much shorter, usually around 6 months to 1 year, so accelerated testing that only takes 3–4 months to produce a full shelf-life prediction is extremely useful. Table 1 details the suggested storage conditions for the accelerated stability test [18].

Table 1. Suggested storage conditions for the accelerated stability test by the International Federation of the Societies of Cosmetic Chemists (IFSCC) [18].

Test Conditions	Time Period of Test
4 °C	projected shelf life of the product
20/25 °C	projected shelf life of the product
37 °C	3–6 months
45 °C	1–3 months
37 °C at 80% relative humidity	1 month maximum

The 4 °C sample is usually used as a control sample as the cold temperature will slow down any changes that may occur. This sample and the 20 °C sample are kept on test for the full shelf life of the product so that a real time stability test can also be carried out to confirm the results of the accelerated stability test. For standard tests 45 °C is generally the maximum temperature as testing samples at higher temperatures such as 50 °C, 60 °C and 70 °C, although theoretically would produce quicker accelerated results, is not recommended for standard testing as stated by the IFSCC, the further removed the conditions are from normal everyday conditions, the more adverse changes happen to the product that likely would never occur in normal conditions. However, testing at these temperatures can be a good initial indicator of the stability of a product, if a sample is still stable at high temperatures such as 60 °C and 70 °C then you can be confident the product will be very stable at normal conditions [12].

Samples need to be observed regularly throughout the testing period for any changes that may occur. The IFSCC recommend that samples are checked frequently at the beginning of a test, generally monthly, but as the test progresses past the first 3 months testing intervals can be spaced further apart. They also suggest that multiple samples of a product should be placed in each condition, enough for each test, so that a new unopened sample is removed from the condition and tested at each interval. When the samples are tested, they should be checked against a control sample for any changes in the appearance, odour, texture, viscosity and pH, the weight loss can also be checked in packaging compatibility samples. Specific parameters will need to be set for each of these properties before testing starts, as the results will likely fluctuate, but the parameters will outline a range that the product is still safe for use and stable between. Table 2 details further testing that may be necessary depending on product or formulation type [18].

Table 2. Additional stability tests that may be necessary according to the IFSCC [18].

	Additional Testing
Cycling Test	Samples that are subjected to a regular change in temperature or humidity can reveal instability quicker than samples stored continuously in one condition.
Freeze Thaw Test	Samples are cycled between −30 °C and room temperature for a minimum of 6 cycles. This can identify formulations that are prone to instability, crystallisation, sedimentation and clouding.
Light/UV Exposure Test	Any product that is likely to be exposed to light in the market should undergo light testing. Samples should be exposed to north facing daylight or placed in a light testing cabinet.
Mechanical Test	Vibration tests or centrifugation can be a good immediate indicator of emulsion stability.

These types of tests are usually performed on products that may face certain conditions in the market or if a formulation or product is known to have past instability issues. Specifically, cycling and freeze–thaw tests are useful when products are planned to be shipped internationally, as they could be subjected to varying extreme temperatures during transport, so these tests will ensure that the products can withstand the extreme temperature changes.

4. Modern Advances in Stability Testing

The demand for innovative cosmetic products that meet current trends is at an all-time high. Competition between cosmetic brands to unveil pioneering concepts is fierce, with indie labels growing rapidly to contend with the majority market shareholders. Online purchasing is enabling consumers to access more beauty brands than ever before, and purchasers no longer have to depend on the limited offering of their local store [21].

Consequently, brands now feel more pressure to quickly capitalise on opportunities in the market whilst the trend is still booming. Standard product development timings consisting of thorough evaluations of product stability are no longer acceptable and are now often ruthlessly reduced to meet the expectations of rapid, ground-breaking beauty launches.

Cutting the time allowed to complete accelerated stability testing is risky. Postles established in a review of the current stability testing guidance that the techniques involved in accelerated stability testing are generally unreliable, concluding that the methodology is inappropriate for predicting long-term shelf life [22]. This deduction was particularly directed towards emulsion technology that founds the basis for many cosmetic formulations. Postles highlighted some potential areas of improvement to the current guidelines in the 2018 study. One recommendation was to encourage the often-overlooked method of real time stability testing to accompany the accelerated testing data. This is indorsed to confirm that the changes observed during the initial program were accurate and to investigate if further instabilities could be expected. It is suggested that real time testing is conducted 6 to 12 months ahead of industrial scale up so that brands can react with speed to any unforeseen changes that may occur when the product is released to market [22]. In the modern market this process may be considered too time-consuming and may add further delay to launches, so is often disregarded in favour of completing post-market surveillance to track success of a formulation. This begs the question of why an inherently untrustworthy method for testing stability can be condensed to meet fast-fashion principles in the beauty market, and what modern advances in technology can be utilised to strengthen the reliability and accuracy of the accelerated testing method.

An effective approach of developing bespoke cosmetic formulations with consumer perceptible differences is the employment of minimally disruptive formulations (MDFs). The MDF concept was described by O'Lenick and Zhang, and involves manipulating a simple, stable base formulation with low levels (<10%) of different silicone polymers to alter the product aesthetics with little disruption to the base's stability profile. The selection of silicone polymers is able to provide assortments of feel, playtime and gloss to the base formulation, providing the formulator with confidence that different consumer perceivable aesthetics are achievable with reasonably predictable stability testing results [23]. The MDF concept can also be extended to include the introduction of low level, novel active ingredients into the base. The engagement of MDFs by beauty brands can dramatically reduce development and stability testing requirements, as well as substantially reduce costs and time frames, without compromising on quality. This method allows brands to quickly react to consumer demands and reduces the risk of missing out on market opportunities.

Advances in technology can also assist with reviewing formulation stability in a timely manner. UK-based CPI have developed their MicroSTAR technology which is a microfluidic platform for stability prediction. This revolutionary, automated method conducts physical testing in shorter timescales to create data for long-term stability prediction with minimal resource and cost. The utilisation of microfluidics inflicts variations of temperature, flow, pressure and vibrations to mimic the diverse environmental conditions that a product may be subjected to during its shelf life. The platform is also capable of revealing insights and drivers of stability failure, which cannot be detected using traditional stability testing methods. CPI offer this technology to enable their clients to conduct the relevant testing in as little as several hours [24].

An alternative method to early stability prediction is the practice of static multiple light scattering (SMLS). SMLS is a high resolution, optical analysis method that is capable of outlining and measuring the destabilisation characteristics of liquid dispersions. This method is particularly useful for cosmetic emulsions, which are thermodynamically unstable and consequently prone to sedimentation, creaming and flocculation. The testing method is carried out on undiluted samples (dilution can affect the dispersion state) and quantifies the rate of changes in particle size and migration and can provide tangible results long before the changes are perceived by the human eye. This process is much quicker than traditional accelerated stability testing methods and could be employed by formulators to promptly forecast formulation stability with precise results [25,26]. Postles advocates the utilization of modern analytical equipment to strengthen shelf-life prediction reliability, stating methods such as examining zeta potential and controlled centrifugation [22].

5. Changes in the Retail Market

How has the retail market changed over the years and how might this influence product stability protocols? Imogen Matthews considers the flexible approach to logistics in a much-changed market, noting that online retail plays a vital role in keeping consumers supplied with beauty goods. She remarks that essential to the success of any brand is the smooth delivery of products in pristine condition to the retailer, online store or directly into the hands of the consumer [27]. Looking at the direct-to-consumer (DTC) business model Schlesinger, Higgins, and Roseman noted that for decades across many categories, including beauty, a handful of brands dominated the consumer retail market. They considered that with the rise of the internet and social media platforms came the rise of the DTC model but concluded that ultimately an omnichannel approach is favourable for growth [28]. This conclusion is supported by a survey McKinsey carried out in 2019 reviewing shopping habits by age group across the cosmetic and skin-care product sectors. They found that while the baby boomer generation had a strong preference to browse and buy in store, generation Z had no strong preference for one shopping habit type, instead preferring an omnichannel approach [29,30].

We need also to consider the impact of the COVID-19 pandemic on consumer shopping habits. By June 2020 Cosmetics Business revealed key insights in a COVID-19 strategy report. They stated that the pandemic has placed unprecedented challenges on the beauty and personal care industry, leaving no brand or retailer untouched. They cited data from McKinsey that estimates a decline of between 20–30% for global beauty industry revenues in 2020. Their report considered that the difficulties experienced by bricks and mortar retailers prior to the pandemic were accelerated. They surmised that as physical stores reopen brands and stores would need to develop new ways for consumers to discover their products [31]. This train of thought is echoed by Lanteri, she notes that for a long time, selling skincare products had largely revolved around brands providing interactive experiences in store [32]. In addition, McKinsey notes that pre-COVID-19 sales in store accounted for up to 85% of beauty purchases in most beauty industry markets, superseding the appeal of shopping online [30]. However, the pandemic has accelerated the shift to online for a much broader section of consumers. Culliney considers this too concluding that beauty brands and retailers must blur physical retail with digital experiences to engage consumers in a post-pandemic world [33]. This opinion is shared by McKinsey in their considerations of the long-term impact of COVID-19 on the beauty industry. They conclude that some changes are likely to be permanent, most notably the rise of digital platforms and the pace of innovations. Their surveys show that across the globe, consumers indicate that they are likely to be increasing their online engagement and spending. They conclude from this that brands will need to prioritise digital channels and overhaul their product innovation pipelines to capitalise on this shift in consumer behaviour [30].

6. Conclusions

Accelerated stability testing is not a precise science, rather a prediction of shelf life. The varied retail approach coupled with the need for pristine product condition highlights the challenges involved in designing a suitable stability protocol. Considering the DTC model, it is important to note that brands are able to reach a global audience where previously they might have only retailed in bricks in mortar in their immediate territory. It is therefore clear that the protocol must consider various modes of transport and sizes of shipments to potentially global destinations. When a container vessel became stuck in the Suez Canal in March 2021 this highlighted the plight of global shipping. Whilst we may not be able to design a protocol that can account for our product been delayed due to this type of force majeure event we must consider that containers can be delayed in global shipping and as a result can be exposed to extremes of temperature. In addition, new and emerging ways of consumer product interaction must be taken into account. The product must remain in specification, being safe, fit for purpose, efficacious and acceptable for the consumer for the duration of the shelf life in order to preserve brand reputation. Stability testing is essential in evaluating a product shelf life and yet it is not set out in regulation how the test should be performed. Global cosmetic associations IFSCC and Cosmetics Europe have published suggested protocols however these were in 1992 and 2004, respectively. The more recent work of Postles encourages the use of real-time stability testing, however this is suggested to be 6–12 months ahead of an industrial scale up which is at odds with a brands desire to move quickly to market. Traditional accelerated testing could be accompanied by additional tests such as cycling, freeze–thaw and centrifuging. Where speed to market is critical employing the formulating approach of minimally disruptive formulations de-risks the process. The advantage that the DTC model has is that brands are not bound by the shelf-life requirements of a retailer and so have the option to consider a shorter shelf life for launch. Post market surveillance can continue after launch and then the shelf life can be gradually extended once more data are gathered to support this. In conclusion to address the diverse market conditions in the context of stability and shelf life, a diverse set of protocols with the need for more product specific stability strategies would seem to be the most logical approach.

Author Contributions: Conceptualisation, L.K., L.H., P.K., H.C.; investigation, L.K., L.H., P.K., H.C.; data curation, L.K., L.H., P.K., H.C.; writing—original draft preparation, L.K., L.H., P.K., H.C.; writing—review and editing, L.K., L.H., P.K., H.C. All authors have read and agreed to the published version of the manuscript.

Funding: This research received no external funding.

Institutional Review Board Statement: Not applicable.

Informed Consent Statement: Not applicable.

Data Availability Statement: Not applicable.

Conflicts of Interest: The authors declare no conflict of interest.

References

1. Dao, H.; Lakhani, P.; Police, A.; Kallakunta, V.; Ajjarapu, S.S.; Wu, K.W.; Ponkshe, P.; Repka, M.A.; Murthy, N. Microbial Stability of Pharmaceutical and Cosmetic Products. Springer Link. 2018. Available online: https://link.springer.com/article/10.1208/s12249-017-0875-1 (accessed on 16 June 2021).
2. Amberg, N.; Fogarassy, C. Green Consumer Behavior in the Cosmetics Market. *Resources* **2019**, *8*, 137. [CrossRef]
3. Verma, A.; Singh, S.; Lohani, A.; Mishra, A.K. Formulation and Evaluation of Carrot Seed Oil-Based Cosmetic Emulsions. Research Gate. May 2018. Available online: https://www.researchgate.net/publication/324890556_Formulation_and_evaluation_of_carrot_seed_oil-based_cosmetic_emulsions (accessed on 16 June 2021).
4. Cambridge Dictionary. Stability. Cambridge Dictionary. Available online: https://dictionary.cambridge.org/dictionary/english/stability (accessed on 9 May 2021).
5. Cosmetics Business. Stability of Cosmetic Products: Shelf Life or PAO? Available online: https://cosmeticsbusiness.com/news/article_page/Stability_of_cosmetic_products_shelf_life_or_PAO/128683 (accessed on 9 May 2021).
6. Halla, N.; Fernandes, I.P.; Heleno, S.A.; Costa, P.; Boucherit-Otmani, Z.; Boucherit, K.; Rodrigues, A.E.; Ferreira, I.C.F.R.; Baerreiro, M.F. Cosmetics Preservation: A Review on Present Strategies. *Molecules* **2018**, *23*, 1571. [CrossRef] [PubMed]
7. European Union. *Regulation (EC) No 1223/2009 of the European Parliament and of the Council*; Official Journal of the European Union, 2009; Available online: https://eur-lex.europa.eu/legal-content/EN/TXT/PDF/?uri=CELEX:32009R1223&from=en (accessed on 9 May 2021).
8. Legislation.gov.uk. The Product Safety and Metrology etc. (Amendment etc.) (EU Exit) Regulations 2019. 2019. Available online: https://www.legislation.gov.uk/ukdsi/2019/9780111180402/contents (accessed on 9 May 2021).
9. U.S. House of Representatives. United States Code Title 21—Food and Drugs, Chapter 9—Federal Food, Drug and Cosmetic Act. Office of the Law Revision Council-United States Code. Available online: https://uscode.house.gov/browse/prelim@title21&edition=prelim (accessed on 16 May 2021).
10. U.S. Food & Drug Administration. Shelf Life and Expiration Dating of Cosmetics. 24 August 2020. Available online: https://www.fda.gov/cosmetics/cosmetics-labeling/shelf-life-and-expiration-dating-cosmetics (accessed on 9 May 2021).
11. Electronic Code of Federal Regulations. United States Code Title 21—Food and Drugs, Chapter I, Subchapter C, Part 211. Electronic Code of Federal Regulations. 13 May 2021. Available online: https://www.ecfr.gov/ (accessed on 16 May 2021).
12. FDA U.S. Food & Drug Administration. Q1A(R2) Stability Testing of New Drug Substances and Products November 2003. FDA U.S. Food & Drug Administration, 24 August 2018. Available online: https://www.fda.gov/regulatory-information/search-fda-guidance-documents/q1ar2-stability-testing-new-drug-substances-and-products (accessed on 16 May 2021).

13. Government of Canada. Guidance for Industry: Stability Testing of New Drug Substances and Products: ICH Topic Q1A(R2). Government of Canada, 25 September 2003. Available online: https://www.canada.ca/en/health-canada/services/drugs-health-products/drug-products/applications-submissions/guidance-documents/international-conference-harmonisation/quality/stability-testing-new-drug-substances-products-topic.html (accessed on 16 May 2021).
14. ASEAN. ASEAN Guidelines for Safety Evaluation of Cosmetic Products. Available online: https://www.fda.gov.ph/wp-content/uploads/2021/03/Guidelines-for-the-Safety-Assessment.pdf (accessed on 16 May 2021).
15. ASEAN. ASEAN Cosmetic Directive Guidelines for Product Information File (PIF). ASEAN, 13 June 2007. Available online: https://www.fda.gov.ph/wp-content/uploads/2021/03/Guidelines_Product-Information-File.pdf (accessed on 16 May 2021).
16. British Standards Institution. *Cosmetics—Guidelines on the Stability Testing of Cosmetic Products*; ISO/TR 1881:2018; BSI Standards Limited: Geneva, Switzerland, 2018.
17. International Federation of Societies of Cosmetic Chemists. Who We Are. IFSCC, 2021. Available online: https://ifscc.org/about/who-we-are/ (accessed on 15 April 2021).
18. International Federation of Societies of Cosmetic Chemists. *IFSCC Monograph Number 2—The Fundamentals of Stability Testing*; Michelle Press: Weymouth, UK, 1992.
19. Cosmetics Europe. About Us. Cosmetics Europe. Available online: https://cosmeticseurope.eu/about-us/ (accessed on 16 May 2021).
20. Cosmetics Europe. Guidelines on Stability Testing of Cosmetic Products. Cosmetics Europe, March 2004. Available online: https://www.cosmeticseurope.eu/files/5914/6407/8121/Guidelines_on_Stability_Testing_of_Cosmetics_CE-CTFA_-_2004.pdf (accessed on 16 May 2021).
21. Dover, S. What's the Future of Independent Beauty Brands? *Mintel Blog*. 11 June 2020. Available online: mintel.com/blog/beauty-market-news/whats-the-future-of-independent-beauty-brands (accessed on 9 May 2021).
22. Postles, A. Factors Affecting the Measurement of Stability and Safety of Cosmetic Products. Ph.D. Thesis, Bournemouth Univeristy, Bournemouth, UK, 2018.
23. O'Lenick, A.J.; Zhang, F. Developing Minimally Disruptive Formulations. Research Gate. April 2015. Available online: https://www.researchgate.net/publication/329152975_Developing_minimally_disruptive_formulations (accessed on 4 May 2021).
24. CPI. Liquid Stability Testing—Accelerating Innovation in High-Throughput Liquid Stability Testing. CPI, 15 January 2021. Available online: https://www.uk-cpi.com/case-studies/high-throughput-stability-testing-platform-for-liquid-formulations (accessed on 4 May 2021).
25. Formulaction. Cosmetic emulsions: Next Generation Shelf-Life Study Towards Formulation Stability Prediction. Formulaction, 24 July 2019. Available online: https://www.formulaction.com/en/about-us/news/blog/formulation-stability-prediction (accessed on 4 May 2021).
26. Formulaction. Static Multiple Light Scattering (SMLS). Formulaction. Available online: https://www.formulaction.com/en/products-and-technologies/technologies/static-multiple-light-scattering-s-mls (accessed on 4 May 2021).
27. Cosmetics Business. Online Logistics: How the Industry's Unsung Heroes Deliver the Goods. Cosmetics Business, 6 May 2020. Available online: https://cosmeticsbusiness.com/news/article_page/Online_logistics_How_the_industrys_unsung_heroes_deliver_the_goods/164773 (accessed on 16 May 2021).
28. Schlesinger, L.; Higgins, M.; Roseman, S. Reinventing the Direct-to-Consumer Business Model. *Harvard Business Review*. 31 March 2020. Available online: https://hbr.org/2020/03/reinventing-the-direct-to-consumer-business-model (accessed on 16 May 2021).
29. Galadari, H.; Gupta, A.; Kroumpouzos, G.; Kassir, M.; Rudnicka, L.; Lotti, T.; Berg, R.V.; Goldust, M. COVID 19 and its Impact on Cosmetic Dermatology. *Dermathol. Thel.* **2020**, *33*, e13822. [CrossRef] [PubMed]
30. McKinsey & Company. How COVID-19 is Changing the World of Beauty. McKinsey & Company, 5 May 2020. Available online: https://www.mckinsey.com/industries/consumer-packaged-goods/our-insights/how-covid-19-is-changing-the-world-of-beauty# (accessed on 16 May 2021).
31. Cosmetics Business. Cosmetics Business Reveals 5 Key Insights in New Covid-19 Strategy Report. Cosmetics Business, 17 June 2020. Available online: https://cosmeticsbusiness.com/news/article_page/Cosmetics_Business_reveals_5_key_insights_in_new_Covid-19_strategy_report/166575 (accessed on 16 May 2021).
32. Lanteri, S. How to Engage Consumers Now: Insights for Skincare Brands. Global Web Index. 10 June 2020. Available online: https://blog.gwi.com/marketing/insights-for-skincare-brands/ (accessed on 16 May 2021).
33. Culliney, K. The Physical-Digital Blur is 'an Absolute Given' for Beauty in 2020, Says Retail Expert. Cosmetics Design Europe. 17 September 2020. Available online: https://www.cosmeticsdesign-europe.com/Article/2020/09/17/Beauty-retail-2020-must-blur-physical-and-digital-to-drive-engagement-in-a-post-COVID-world?utm_source=copyright&utm_medium=OnSite&utm_campaign=copyright (accessed on 16 May 2021).

Perspective

Fractions of Concern: Challenges and Strategies for the Safety Assessment of Biological Matter in Cosmetics

Fabian P. Steinmetz [1], James C. Wakefield [2] and Ray M. Boughton [3,*]

1. Delphic HSE (Europe) B.V., 1118CN Schiphol, The Netherlands; fabian.steinmetz@delphichse.com
2. Delphic HSE Solutions (HK) Ltd., Shatin, Hong Kong, China; james.wakefield@delphichse.com
3. Delphic HSE Solutions Ltd., Camberley GU15 3YL, UK
* Correspondence: ray.boughton@delphichse.com

Abstract: Cosmetic ingredients based on more or less refined biological matter (plants, fungi, bacteria, etc.) are gaining popularity. Advances in green chemistry and biotechnology are supporting this general trend further. Following numerous bans on the use of newly generated animal testing data in cosmetic safety assessments, and the worldwide demand for "cruelty-free" products, many alternative methods have been developed to assess the toxicity of ingredients. Whilst great strides have been, and continue to be, made, the area of systemic toxicity is one where international harmonisation and regulatory acceptance is still evolving. A strategy for the fractional assessment of biological matter is suggested to make approaches, such as threshold of toxicological concern (TTC) methodology, fit for purpose. Within this strategy, analytical data are used to generate compound classes which are quantified and assessed separately. Whilst this strategy opens new windows for assessing the safety of complex mixtures with a lack of toxicological data, it also raises awareness of the increasing complexity of cosmetic formulations and the general problem of additivity/synergy being rarely addressed. Extremely complex mixtures are and will be a growing challenge for safety assessors.

Keywords: safety; cosmetics; botanicals; toxicology; TTC

1. Introduction

Cosmetic products may contain a plethora of chemical compounds which themselves may originate from different sources. These sources refer to biological matter from plants, fungi, animals, bacteria and algae but also mineral matter, such as fractions from mineral oil or pigments, and of course derivatives and combinations due to chemical reactions and biotechnological processes. As a rule of thumb, the less purification of educts and products is conducted, the more complex the chemistry of the resulting ingredient. However, public perspective, international trade and advances in green chemistry and biotechnology are creating a shift towards more biological matter from plants, fungi, algae and bacteria as ingredients for consumer goods, such as cosmetics [1–3]. Although biological ingredients may be perceived as natural and safer by consumers, complex chemical mixtures are difficult to assess, independent from their origin. In general, most plant extracts are complex mixtures and prone to a certain variability based on season, utilised plant parts and solvents but also process parameters, such as temperature and pressure. It is easy to imagine how difficult safety assessments might become when products are assessed containing multiple botanicals and related materials. Furthermore, such challenges did not become easier by voluntary and mandatory animal testing bans, because these are limiting toxicological testing batteries.

There is a worldwide shift to "cruelty-free" cosmetic products, for example manifested in Regulation (EC) No. 1223/2009 [4] that banned animal testing in the EU for cosmetic products and ingredients. Nevertheless, the safety of the ingredients used in a cosmetic product is a key feature for the safety of cosmetics as described in SCCS/1602/18 [5].

Historical data from animal testing have been used by safety assessors to exclude the risks of significant skin and eye irritation, skin sensitisation, genotoxicity and systemic toxicity. In light of the animal testing bans for cosmetics, most notably in the EU, and the general drive to move away from risk assessments based on animal test data, alternative approaches are required. Whilst there are nowadays many alternative testing methods available, e.g., bacterial mutation test (Ames test), Hen's egg test on chorioallantoic membrane (HET-CAM) assay, bovine corneal opacity and permeability (BCOP) assay, direct peptide reactivity assay (DPRA), and human cell line activation test (h-CLAT), systemic toxicity can only be partially addressed, for example, when modes of action (MoA) are known [6]. Hence, current strategies include MoA-driven testing/analysis, investigations on the history of safe use [7], read-across approaches [8] and the threshold of toxicological concern (TTC) methodology [9,10]. It should be mentioned that in a weight-of-evidence (WoE) approach, combinations of those strategies are possible, for example, an on-its-own insufficient read-across could be supported by negative in silico predictions/bioassay results for a potential MoA and/or by limited history of safe use data.

Particularly, the TTC is a popular way to justify safety of biological matter, such as plant extracts or ferments, with regard to systemic toxicity. This manuscript asks critically whether the current TTC approach or derivatives thereof are fit for purpose but also suggests further refinements which allow for more flexibility based on the available data.

2. Threshold Approaches

The current TTC approach is based on "Cramer classes" [11], which itself is a classification system for chemical compounds. The origin of this approach lies in the assessment of low-level substances in the human diet. Basically, there are three different classes which are categorised with increasing toxicological concern. Class I is more associated with endogenous or rather inert compounds, while Class III is more associated with drug-like or reactive (potentially toxic) compounds. Class II fits the spectrum in between those two classes. Many chemoinformatic tools, such as ToxTree v3.1.0 (Ideaconsult Ltd, Brussels, Belgium) [12] or OECD QSAR Toolbox 4.4.1. (OASIS LMC, Burgas, Bulgaria) [13], use this decision tree or derivatives thereof.

Munro and colleagues assigned threshold values for those classes based on 95th percentiles of no-observed-adverse-effect-level (NOAEL) data, Yang and colleagues validated and refined these thresholds with new data in 2017 [9,10]. The resulting thresholds are 46 µg/kg bw/day for Cramer Class I and 2.3 µg/kg bw/day for Cramer Class II and III (with a bodyweight defined as 60 kg). If a structural alert for genotoxicity is triggered, then the threshold should be reduced to 0.0025 µg/kg bw/day according to Kroes and colleagues [14]—this can be considered as an unofficial "fourth Cramer class". Nevertheless, in vitro genotoxicity testing is considered preferable, which limits the necessity for in silico genotoxicity investigations. In Figure 1, as an example, three unrelated but chemically similar compounds were investigated with ToxTree v3.1.0 [12], followed up by the assigning of appropriate classes and thresholds.

Although not the focus of this manuscript, the dermal sensitisation threshold (DST) shall not remain unnamed. Here, a similar approach is applied utilising reactivity domains (cf. structural alerts) and skin sensitisation data [15–18].

It must be emphasised that the TTC approach is intended for individual compounds and not for mixtures, and that assigning 2.3 µg/kg for all biological matter (after genotoxicity was excluded via in vitro testing) might be considered overly conservative, i.e., safe products with low to moderate exposure might fail a safety assessment. With regard to TTC for biological matter, Kawamoto and colleagues [19] suggested to either use a Cramer Class III threshold for botanicals (which was found protective) or to use the 1st percentile of their data analysis: 663 µg/day or 11.05 µg/kg bw/day (bodyweight defined as 60 kg). Both approaches are rather conservative and try to comprise the huge variety in toxicity. Biological raw materials are complex mixtures with a large chemical variability and therefore a one-size-fits-all approach might become overly conservative for many raw materials of interest.

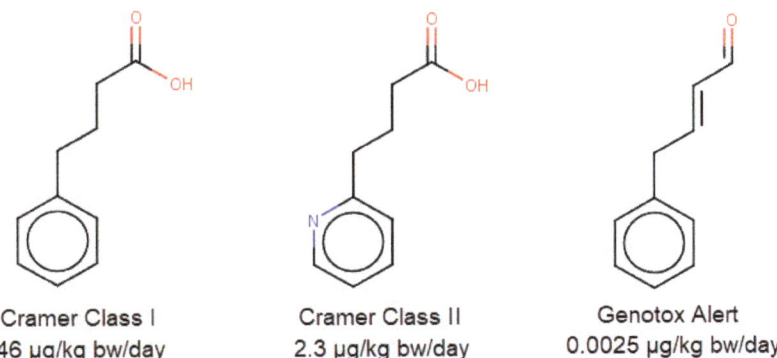

Figure 1. Three structurally similar compounds investigated via ToxTree v3.1.0. From left to right: 4-phenylbutyric acid (CAS 1821-12-1), 4-(2-pyridinyl) butanoic acid (CAS 102879-51-6) and 4-phenylcrotonaldehyde (CAS 13910-23-1).

3. Fractions of Concern

As biological matter may contain a huge variety of chemical compounds, it is best to either conduct a literature search for relevant analytical data or to perform a chemical analysis for the material of interest. Specifications and certificates of analysis may help with standardisation, so chemical data are transferable between batches and raw material suppliers. In case this is not possible, information on taxonomy, plant parts, solvents/processes used, etc., may help with approaches to overcome data gaps/uncertainties. In general, conservative estimations/safety buffers are recommended for such approaches. Furthermore, genotoxicity should be excluded by in vitro testing, so the classification may focus around the three Cramer classes.

Biological materials, such as an ethanol/water extract of the aerial parts of a common herb, must be considered as a complex mixture of phytochemicals, e.g., chlorophyll, tannins, alkaloids, fatty acids, amino acids, sugars and terpenoids. Many of these compounds can be regarded as Cramer Class I, i.e., endogenous or rather inert and consequently of low concern. Alkaloids or specific tannins and terpenoids, for instance, might be of higher concern (cf. Cramer Class II and III). When splitting the systemic exposure dose (SED) quantitatively according to Cramer Class I compounds and Class II and III compounds, the margin of safety (MoS) calculation can be executed separately (cf. Equations (1) and (2) with regard to Cramer Class I and Cramer Class II + III, respectively).

$$MoS_{fract\ I} = \frac{46\ \mu g/kg\ bw/day}{SED_{fract\ I}} \qquad (1)$$

$$MoS_{fract\ II+III} = \frac{2.3\ \mu g/kg\ bw/day}{SED_{fract\ II+III}} \qquad (2)$$

Apart from TTC values, points of departure, such as NOAEL and acceptable daily intake (ADI) can be used if toxicological data are available. Furthermore, rationales, such as history of safe use, can be used for risk assessing. Principally, if the SED is higher than the TTC or an alternative point of departure (with appropriate safety factors), then this must be considered a violation suggesting a lack of safety. The overall scheme of this approach is expressed in Figure 2.

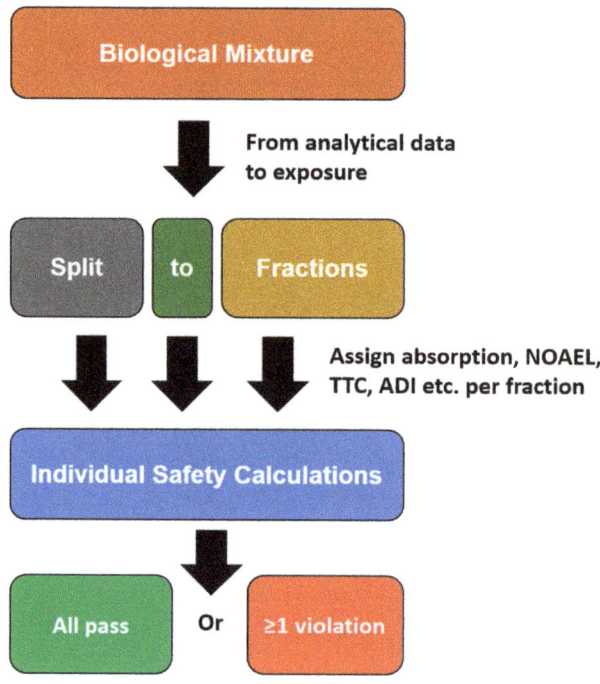

Figure 2. Fractions of concern scheme.

If all fractions are considered safe (i.e., there are no violations present), then the incorporated biological matter is considered safe. However, as chemical information is often not that detailed, there is some flexibility with regard to the definitions of fractions. An example for a fraction could for instance be simple fatty acids and their esters. Beyond this flexibility in the definition of fractions, there is also some flexibility in arguments with regard to safety in both directions, i.e., either arguing for safety and arguing for lack of safety-relevant information. While low dermal absorption, rapid metabolism or information regarding safe use may lead to some tolerance in case of mild MoS violations, aggregated exposure, synergy/additivity towards other ingredients or insufficient analytical data may suggest a MoS even lower than mathematically expressed for the individual raw material. Overall, this approach still entails some degrees of freedom for the assessor.

As mentioned above, interpreting analytical data and defining fractions are probably the most challenging parts. For instance, the composition of *Camellia sinensis* in the form of green tea, black tea and infusions thereof were described by Chacko and colleagues [20]. However, as the description is rather crudely classified into amino acids, minerals, polyphenols, etc., no chemical data on potentially active polyphenols were provided. Here, the publication of Reto and colleagues [21] might help to identify key components which then can be used for a toxicological literature review or they can be used for an investigation with an appropriate chemoinformatic tool to assign Cramer classes (e.g., ToxTree v3.1.0 [12] or OECD QSAR Toolbox 4.4.1. [13]). While there are many aqueous extracts of green tea being used in cosmetics [22], some may be more concentrated extracts as compared to a simple infusion or even different extraction solvents and process parameters being used. These may influence the final composition of the extract significantly.

Ideally a fractional process would be conducted for a complete cosmetic formulation to address potential additivity/synergy, at least for an obvious MoA, e.g., retinoid-like compounds (cf. vitamin A esters) and compounds with an estrogenic potential, such as parabens, 2-ethylhexanoate and certain steroids [23–26].

4. Perspective

Assessing the safety of biological matter and the complex mixtures they may entail is not a trivial matter. Voluntary and mandatory bans on animal testing demand novel solutions addressing systemic toxicity [6,27]. Such novel solutions to assess safety may include fractional approaches which can either be used stand-alone or as part of a WoE approach. However, due to the complexity of cosmetic formulations often containing multiple extracts/ferments, it is important to initially obtain a grasp of the relevant chemistry. Particularly for a toxicologically relevant MoA, chemical compounds from mixtures must also be considered towards additivity/synergy and those fractions must be addressed in appropriate safety calculations. Neither the complexity of a mixture nor testing restrictions are reasons for insufficient risk assessments. Despite all restraints, consumer safety is key.

Author Contributions: All authors have read and agreed to the published version of the manuscript.

Funding: This research received no external funding.

Institutional Review Board Statement: Not applicable.

Informed Consent Statement: Not applicable.

Data Availability Statement: Not applicable.

Conflicts of Interest: The authors declare no conflict of interest.

References

1. Rinaldi, A. Healing beauty? More biotechnology cosmetic products that claim drug-like properties reach the market. *EMBO Rep.* **2008**, *9*, 1073–1077. [CrossRef]
2. Allemann, I.B.; Bauman, L. Botanicals in skin care products. *Int. J. Dermatol.* **2009**, *48*, 923–934. [CrossRef]
3. Becker, M.; Tickner, J.A. Driving safer products through collaborative innovation: Lessons learned from the Green Chemistry & Commerce Council's collaborative innovation challenge for safe and effective preservatives for consumer products. *Sustain. Chem. Pharm.* **2020**, *18*, 100330.
4. Regulation (EC) No. 1223/2009 of the European Parliament and of the Council of 30 November 2009 on cosmetic products (recast). *Off. J. Eur. Union Luxemb.* **2009**, *342*, 59–209.
5. Bernauer, U.; Bodin, L.; Chaudry, Q.; Coenraads, P.J.; Dusinka, M.; Ezendam, J.; Gaffet, E.; Galli, C.L.; Granum, B.; Rogiers, V.; et al. *The SCCS Notes of Guidance for the Testing of Cosmetic Ingredients and Their Safety Evaluation*, 10th ed.; SCCS: Brussels, Belgium, 2019.
6. Díez-Sales, O.; Nácher, A.; Merino, M.; Merino, V. *Chapter 17-Alternative Methods to Animal Testing in Safety Evaluation of Cosmetic Products. Analysis of Cosmetic Products*, 2nd ed.; Salvador, A., Chisvert, A., Eds.; Elsevier: Amsterdam, The Netherlands, 2018; pp. 551–584.
7. Constable, A.; Jonas, D.; Cockburn, A.; Davi, A.; Edwards, G.; Hepburn, P.; Herouet-Guicheney, C.; Knowles, M.; Moseley, B.; Oberdörfer, R.; et al. History of safe use as applied to the safety assessment of novel foods and foods derived from genetically modified organisms. *Food Chem. Toxicol.* **2007**, *45*, 2513–2525. [CrossRef]
8. ECHA. *Read-Across Assessment Framework (RAAF)*; ECHA: Helsinki, Finland, 2017. [CrossRef]
9. Munro, I.C.; Ford, R.A.; Kennepohl, E.; Sprenger, J.G. Correlation of a structural class with No-Observed-Effect-Levels: A proposal for establishing a threshold of concern. *Food Chem. Toxicol.* **1996**, *34*, 829–867. [CrossRef]
10. Yang, C.; Barlow, S.M.; Jacobs, K.L.M.; Vitcheva, V.; Boobis, A.R.; Felter, S.P.; Arvidson, K.B.; Keller, D.; Cronin, M.T.D.; Enoch, S.; et al. Thresholds of Toxicological Concern for cosmetics-related substances: New database, thresholds, and enrichment of chemical space. *Food Chem. Toxicol.* **2017**, *109*, 170–193. [CrossRef] [PubMed]
11. Cramer, G.M.; Ford, R.A.; Hall, R.L. Estimation of toxic hazard-a decision tree approach. *Food Cosmet. Toxicol.* **1978**, *16*, 255–276. [CrossRef]
12. ToxTree v3.1.0. Toxic Hazard Estimation by Decision Tree Approach. Available online: http://toxtree.sourceforge.net/ (accessed on 11 November 2018).
13. OECD 4.4.1. OECD QSAR Toolbox 4.4.1. Available online: https://qsartoolbox.org/download/ (accessed on 11 January 2021).
14. Kroes, R.; Renwick, A.G.; Cheeseman, M.; Kleiner, J.; Mangelsdorf, I.; Piersma, A.; Schilter, B.; Schlatter, J.; van Schothorst, F.; Vos, J.G.; et al. Structure-based thresholds of toxicological concern (TTC): Guidance for application to substances present at low levels in the diet. *Food Chem. Toxicol.* **2004**, *42*, 65–83. [CrossRef]
15. Payne, M.P.; Walsh, P.T. Structure-activity relationships for skin sensitization potential: Development of structural alerts for use in knowledge-based toxicity prediction systems. *J. Chem. Inf. Comput. Sci.* **1994**, *34*, 154–161. [CrossRef]
16. Roberts, D.W.; Api, A.M.; Safford, R.J.; Lalko, J.F. Principles for identification of High Potency Category Chemicals for which the Dermal Sensitisation Threshold (DST) approach should not be applied. *Regul. Toxicol. Pharmacol.* **2015**, *72*, 683–693. [CrossRef]

17. Safford, R.J.; Aptula, A.O.; Gilmour, N. Refinement of the Dermal Sensitisation Threshold (DST) approach using a larger dataset and incorporating mechanistic chemistry domains. *Regul. Toxicol. Pharmacol.* **2011**, *60*, 218–224. [CrossRef] [PubMed]
18. Safford, R.J.; Api, A.M.; Roberts, D.W.; Lalko, J.F. Extension of the Dermal Sensitisation Threshold (DST) approach to incorporate chemicals classified as reactive. *Regul. Toxicol. Pharmacol.* **2015**, *72*, 694–701. [CrossRef] [PubMed]
19. Kawamoto, T.; Fuchs, A.; Fautz, R.; Morita, O. Threshold of Toxicological Concern (TTC) for Botanical Extracts (Botanical-TTC) derived from a meta-analysis of repeated-dose toxicity studies. *Toxicol. Lett.* **2019**, *316*, 1–9. [CrossRef] [PubMed]
20. Chacko, S.M.; Thambi, P.T.; Kuttan, R.; Nishigaki, I. Beneficial effects of green tea: A literature review. *Chin. Med.* **2010**, *5*, 13. [CrossRef]
21. Reto, M.; Figueira, M.E.; Filipe, H.M.; Almeida, C.M. Chemical composition of green tea (Camellia sinensis) infusions commercialized in Portugal. *Plant. Foods Hum. Nutr.* **2007**, *62*, 139–144. [CrossRef]
22. Prasanth, M.I.; Sivamaruthi, B.S.; Chaiyasut, C.; Tencomnao, T. A Review of the Role of Green Tea (Camellia sinensis) in Antiphotoaging, Stress Resistance, Neuroprotection, and Autophagy. *Nutrients* **2019**, *11*, 474. [CrossRef]
23. SCHER; SCCS; SCENIHR. Opinion on the Toxicity and Assessment of Chemical Mixtures. 2012. Available online: https://ec.europa.eu/health/scientific_committees/environmental_risks/docs/scher_o_155.pdf (accessed on 11 February 2021).
24. Boobis, A.; Budinsky, R.; Collie, S.; Crofton, K.; Embry, M.; Felter, S.; Hertzberg, R.; Kopp, D.; Mihlan, G.; Mumtaz, M.; et al. Critical analysis of literature on low-dose synergy for use in screening chemical mixtures for risk assessment. *Crit. Rev. Toxicol.* **2011**, *41*, 369–383. [CrossRef]
25. Kortenkamp, A.; Faust, M.; Scholze, M.; Backhaus, T. Low-level exposure to multiple chemicals: Reason for human health concerns? *Environ. Health Perspect.* **2007**, *115* (Suppl. 1), 106–114. [CrossRef]
26. Martin, O.; Scholze, M.; Ermler, S.; McPhie, J.; Bopp, S.K.; Kienzler, A.; Parissis, N.; Kortenkamp, A. Ten years of research on synergisms and antagonisms in chemical mixtures: A systematic review and quantitative reappraisal of mixture studies. *Environ. Int.* **2021**, *146*, 106206. [CrossRef]
27. Antignac, E.; Nohynek, G.J.; Re, T.; Clouzeau, J.; Toutain, H. Safety of botanical ingredients in personal care products/cosmetics. Food and chemical toxicology. *Food Chem. Toxicol.* **2011**, *49*, 324–341. [CrossRef] [PubMed]

MDPI
St. Alban-Anlage 66
4052 Basel
Switzerland
Tel. +41 61 683 77 34
Fax +41 61 302 89 18
www.mdpi.com

Cosmetics Editorial Office
E-mail: cosmetics@mdpi.com
www.mdpi.com/journal/cosmetics

www.ingramcontent.com/pod-product-compliance
Lightning Source LLC
LaVergne TN
LVHW070541100526
838202LV00012B/342